NELSON MATHS

VICTORIAN CURRICULUM

3

Student Book

Yale Mercieca and Lauren White

NELSON
A Cengage Company

Contents

Odd or Even?

DATE:

You will need: counters 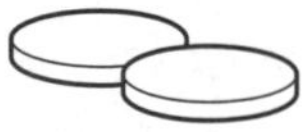

1 Use counters to work out if the following numbers are **odd** or **even**.

1 2 3 4 5 6 7 8 9 10

Colour the **odd** numbers green and the **even** numbers yellow in the chart below.

2 Use counters to work out if the following numbers are **odd** or **even**.

11 12 13 14 15 16 17 18 19 20

Colour the **odd** numbers green and the **even** numbers yellow in the chart below.

3 Is there a pattern? ______________ What is the pattern?

____

4 Use the pattern to colour all the rest of the **odd** numbers green and the **even** numbers yellow.

1	2	3	4	5	6	7	8	9	10
11	12	13	14	15	16	17	18	19	20
21	22	23	24	25	26	27	28	29	30
31	32	33	34	35	36	37	38	39	40
41	42	43	44	45	46	47	48	49	50
51	52	53	54	55	56	57	58	59	60
61	62	63	64	65	66	67	68	69	70
71	72	73	74	75	76	77	78	79	80
81	82	83	84	85	86	87	88	89	90
91	92	93	94	95	96	97	98	99	100

5 What do you notice about all of the odd numbers? ______________

____

6 What do you notice about all of the even numbers? ______________

____

Number Sort

DATE:

1 Write the following numbers in word form in the table below. Make sure you put them in the correct box!

372 621 509 196 765 12 54 440 3007

	Even	Odd
Numbers more than 500		
Numbers less than 500		

2 Write your own 3-digit number here. Write it in words in the table above.

True Statement

DATE:

You will need: a partner, a 10-sided dice 

1 Take turns to roll the dice.

Each time you roll the dice, write the number in any empty box.

Repeat these steps until you have a number in every box.

Score 1 point if you make a statement true.

☐ ☐ ☐ is an even number less than 550.

☐ ☐ ☐ ☐ is greater than 7500.

☐ ☐ ☐ ☐ is a number less than 2500 but greater than 2000.

☐ ☐ ☐ ☐ is an odd number less than 5500.

☐ ☐ ☐ ☐ is a number between 6000 and 7000.

☐ ☐ ☐ ☐ is a number more than 8000.

Total score: ____________

2 Play the game again. See if you can get a better score.

☐ ☐ ☐ is an even number less than 550.

☐ ☐ ☐ ☐ is greater than 7500.

☐ ☐ ☐ ☐ is a number less than 2500 but greater than 2000.

☐ ☐ ☐ ☐ is an odd number less than 5500.

☐ ☐ ☐ ☐ is a number between 6000 and 7000.

☐ ☐ ☐ ☐ is a number more than 8000.

Total score: ____________

3 Which numbers were the hardest to score a point on? Why?

____

Which numbers did you try to fill in first? Why?

____

Numbers, Numbers, Numbers (TRB pp. 22–25)
Number and place value Investigate the conditions required for a number to be odd or even and identify odd and even numbers (VCMNA129)

Recognise, model, represent and order numbers to at least 10 000 (VCMNA130)

Unit 1

DATE:

STUDENT ASSESSMENT

1 Look at the numbers. Draw a circle around the **odd** numbers.

16 57 99 123 312 450 671 3025 7138 9734

How do you know the numbers you circled are odd?

____

2 Write the following numbers in **words**.

a 2763 ______________________________

b 919 ______________________________

c 5044 ______________________________

d 3900 ______________________________

3 Write the following numbers as **numerals**.

a Three thousand, four hundred and fifty-two

____

b Eight thousand and forty-seven

____

c Two hundred and ninety-six

____

d Seven thousand and thirteen

____

4 Use the four digits below to make the following numbers.

1 8 7 6

a an even number **more than** 8500

____

b an odd number **less than** 3500

____

c a number **between** 5000 and 8000

____

4-digit Numbers

DATE:

1 Draw lines matching the numerals, MAB blocks and number words.

2 354

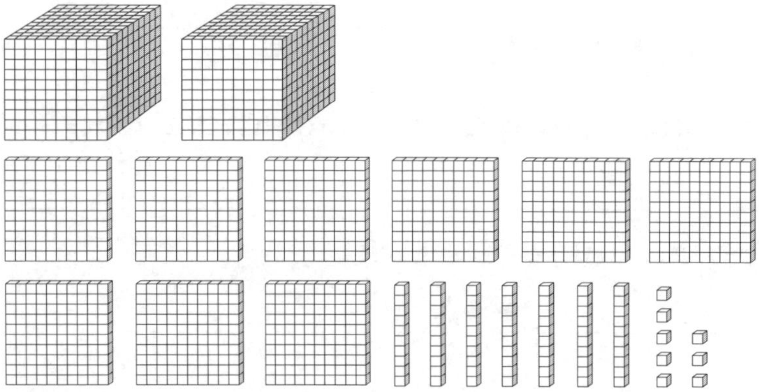

Two thousand, three hundred and fifty-four

2 978

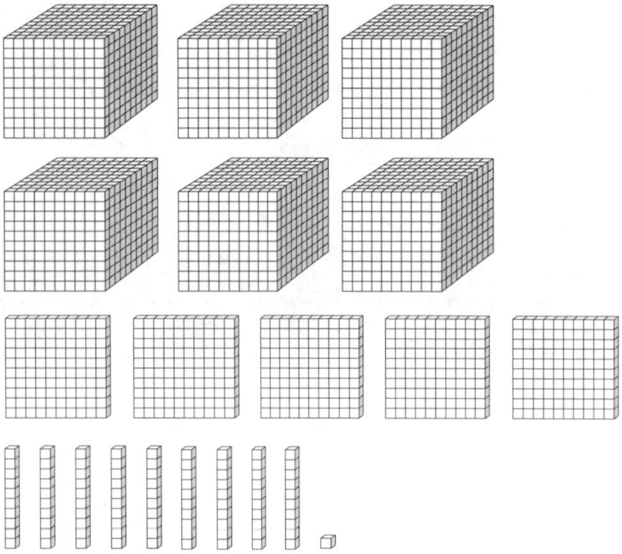

Two thousand, nine hundred and seventy-eight

6 591

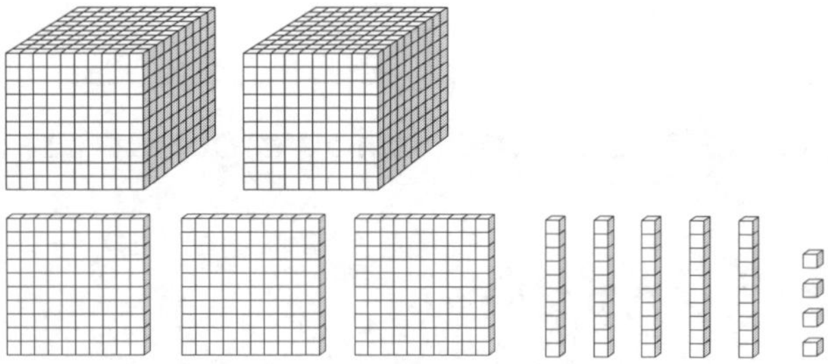

Six thousand, five hundred and ninety-one

2 Write the **largest** number from Question 1. ____________________

3 Write the numbers from Question 1 in order from **smallest** to **largest**.

____

Road Trip!

DATE:

Jill drives a truck and makes deliveries all over Australia. She has to make sure she gets to all of the capital cities on the mainland, and still make her deliveries on time! Here is Jill's travel map.

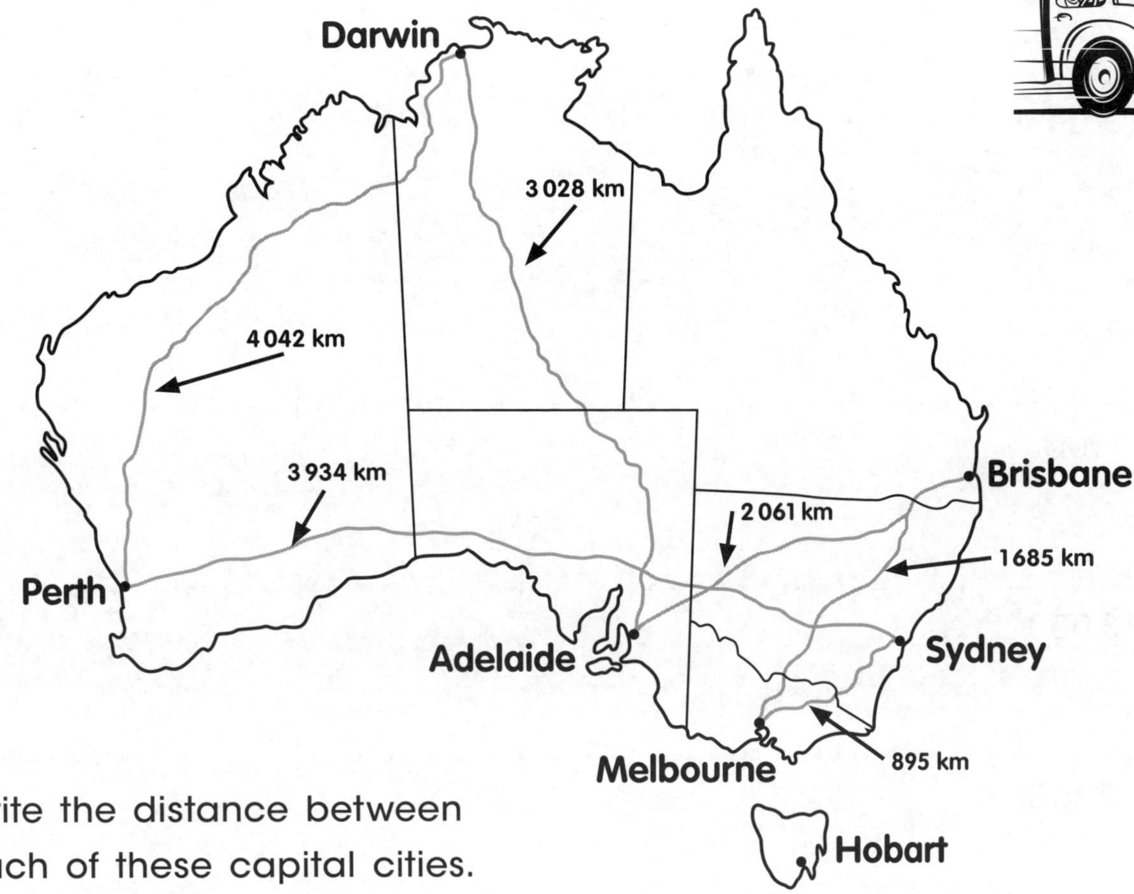

1 Write the distance between each of these capital cities.

Melbourne to Sydney	km
Sydney to Perth	km
Perth to Darwin	km
Darwin to Adelaide	km
Adelaide to Brisbane	km
Brisbane to Melbourne	km

2 Write the distances from **smallest** to **largest**.

3 Can you think of a different route Jill could take to reach all of the capital cities with less driving? If so, draw your idea on the map.

Line Them Up!

DATE:

1 Write these numbers on the correct number lines. Note: you might not use all of the numbers.

102 1 490 4 973 1 510 3 886

1 411 3 986 180 1 490 3 005 165

183 99 3 210 1 560 154 7 894

3 750 1 531 191 3 290 3 190

a 

b 

c 

2 Circle the **largest** number in each pair.

a One hundred and twenty-nine — One hundred and fifty-six

b Two thousand, nine hundred and fifty — Three thousand and one

c Seven thousand and ninety — Seven thousand, one hundred and thirty-five

3 Which number is larger: 2 987 or 2 789? ______________

How do you know?

____

Unit 2

DATE:

STUDENT ASSESSMENT

1 Write the numeral for each MAB model.

a

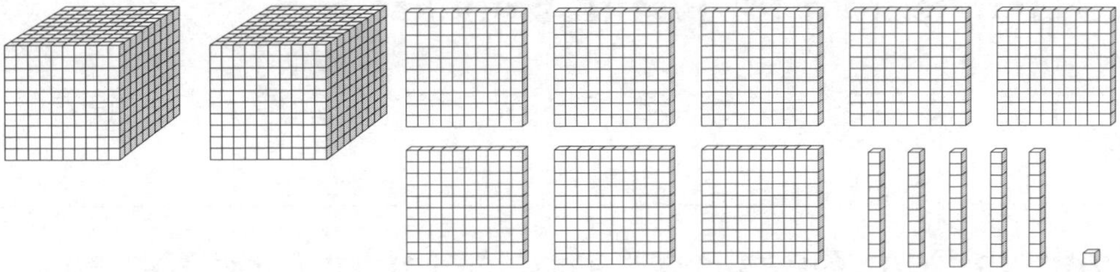

b

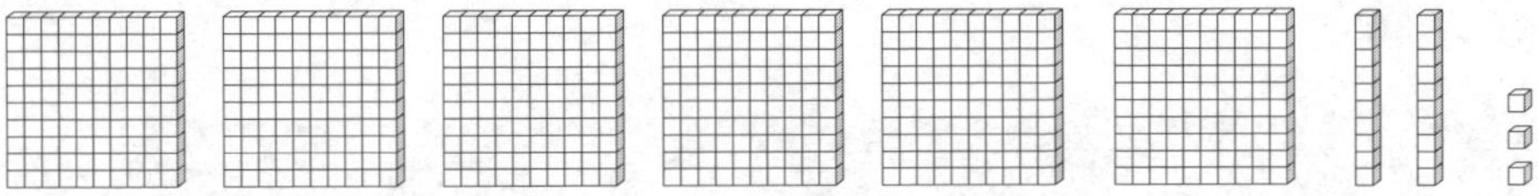

c

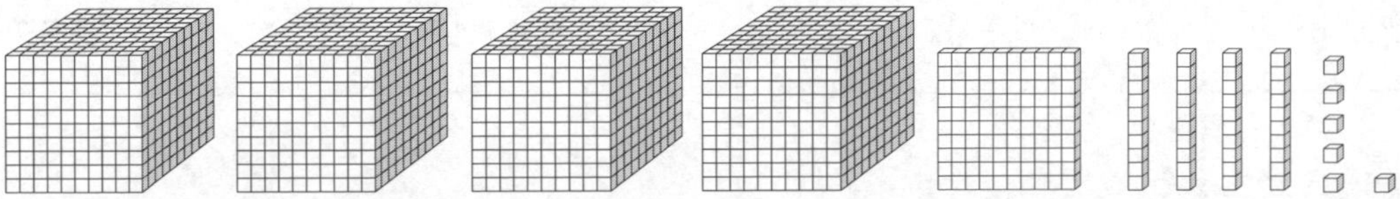

2 How many MAB thousands blocks would you need to make the following numbers?

a 8 231 ____________ **b** 789 ____________ **c** 3 289 ____________

3 Circle the **largest** number in each set.

a 512 673 599

b 5 912 5 891 5 199

c 2 673 1 978 3 415

d 9 812 8 999 9 299

4 Order these numbers from **smallest** to **largest**.

a 3 829 3 672 3 119 3 578 3 511 3 355

____

b How did you know what number was:

- the smallest? ____________________________
- the largest? ____________________________

5 Place the numbers from Question 4 on the number line.

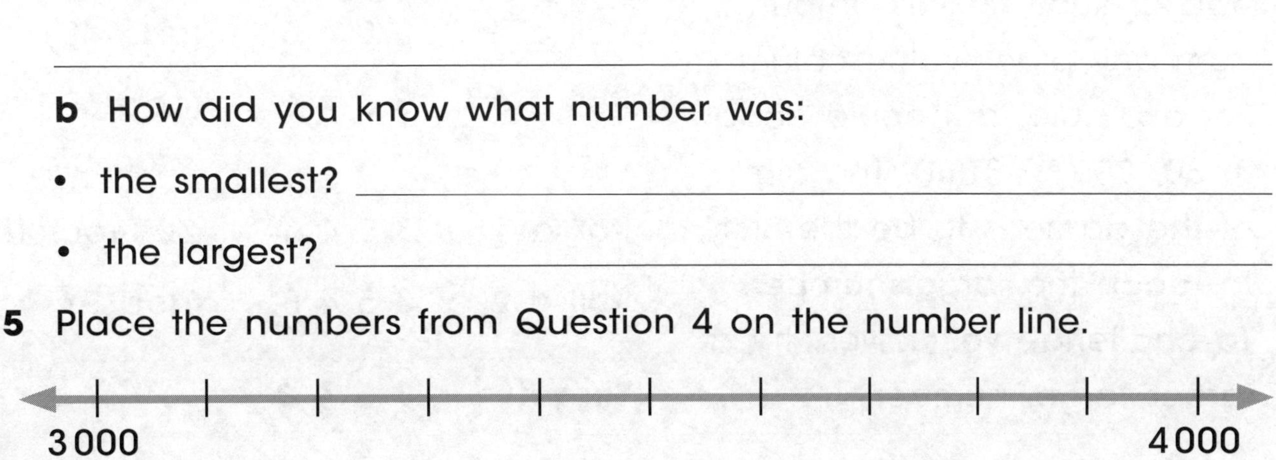

Adding and Subtracting Value

DATE:

You will need: a dice, extra paper for Question 3, a partner

1 Write these numbers from **largest** to **smallest**.

3574 8121 6978 3548 649 7893 956 8119

2 Select 3 numbers from Question 1 and complete the table. One has been done.

Original number	Add 10	Add 100	Add 1000
3574	35**8**4	3**6**74	**4**574

3 With your partner, select one number from Question 1. This is the **target number**. Both players start at 1000. In turn, roll the dice. Whichever number you roll, you can add or subtract this amount from any place-value column. For example, a 3 could equal 3, 30, 300 or 3000. The aim of the game is to be the first to reach the target number. To challenge yourselves, try a larger target number.

See example. Target number: **3548**

Start: **1 0 0 0**

Roll a **2**	3 0 0 0	(add 2000)
Roll a **6**	3 0 0 6	(add 6 – *too small*)
Roll a **4**	3 0 4 6	(add 40)
Roll a **6**	3 6 4 6	(add 600 – *too big*)
Roll a **1**	3 5 4 6	(subtract 100)
Roll a **2**	3 5 4 8	(add 2)

Target: **3 5 4 8**

Birthday Values

DATE:

Write your birthday number and fill in the values box. Then, follow the addition and subtraction path. You may use MAB blocks, a spike abacus or a calculator to help you.

Do this first!

Write the day and month of your birthday as a 4-digit number (e.g. 22 January is 2201)

___ ___ ___ ___

What is the value of the digit in the:

ones column? ________

tens column? ________

hundreds column? ________

thousands column? ________

Add 100.

___ ___ ___ ___

Subtract 10.

___ ___ ___ ___

Add 1000.

___ ___ ___ ___

Subtract 90.

___ ___ ___ ___

Add 300.

___ ___ ___ ___

Subtract 2000.

___ ___ ___ ___

On another sheet of paper, you could make your own "Birthday Values" addition and subtraction paths for a partner to solve.

Numbers on an Abacus

DATE:

1 From the numbers shown on the dice, create the **largest** number you can on the abacus.

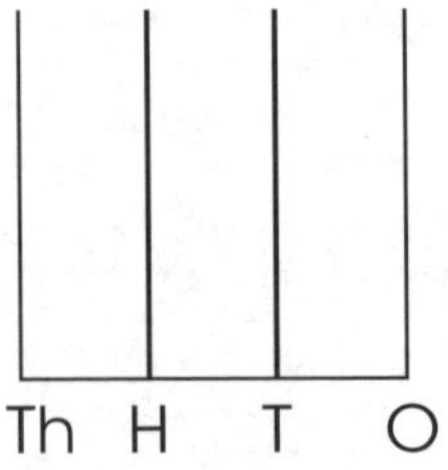

a How many thousands are in this number? ____________

b Write the value of the 3 in this number. ____________

c Add 80 to the number. ____________ + 80 = ____________

d Subtract 3 from the number. ____________ - 3 = ____________

2 From the digits shown on the dice, create the **smallest** number you can.

a How many thousands are in this number? ____________

b Write the value of the 3 in this number. ____________

c Add 30 to the number. ____________ + 30 = ____________

d Subtract 400 from the number.

____________ - 400 = ____________

3 Create 3 other numbers using the digits on the dice. Write each number in a box. Model each number on an abacus, then add or subtract the amount shown and write the new number.

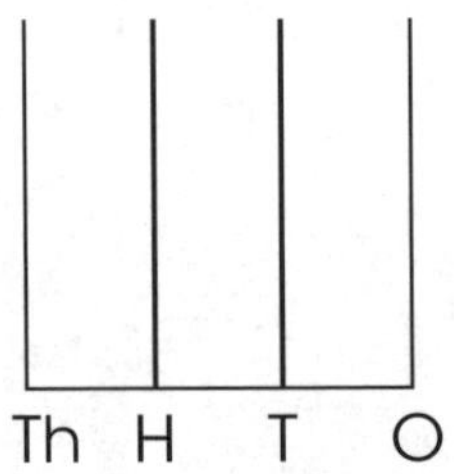

Add 500:

____

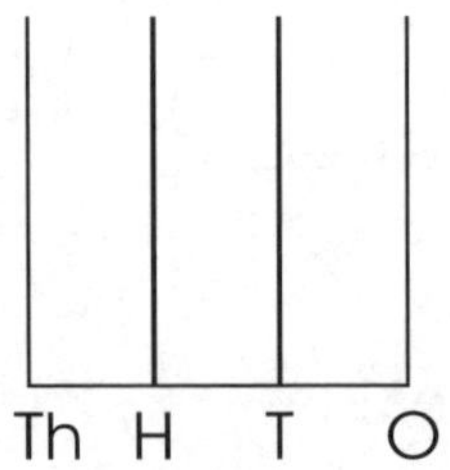

Subtract 2000:

____

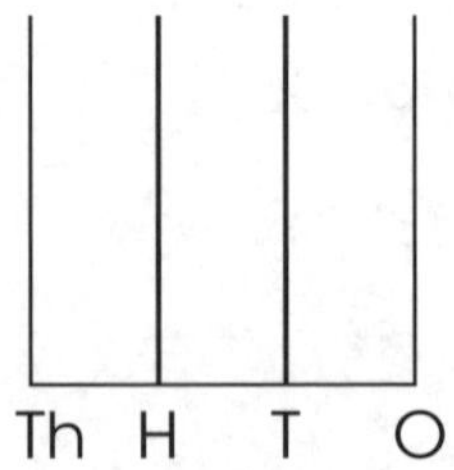

Subtract 80:

____

More About Numbers to 10000 (TRB pp. 30–33)
Number and place value Recognise, model, represent and order numbers to at least 10000 (VCMNA130)

Unit 3

DATE:

STUDENT ASSESSMENT

1 **a** Using the digits 2, 4, 6 and 7, create the **largest number** you can. ______________

b How many thousands are in this number? ______________

c What is the value of the 6? ______________

2 **a** Using the digits 2, 4, 6 and 7, create the **smallest number** you can. ______________

b How many thousands are in this number? ______________

c What is the value of the 6? ______________

3 **a** Calculate **100 more** for each of these numbers.

3 178   5 497   4 219   7 738   5 682   2 941

________ ________ ________ ________ ________ ________

b Explain how you calculated 100 more.

____

4 **a** Calculate **10 less** for each of these numbers.

3 178   5 497   4 219   7 738   5 682   2 941

________ ________ ________ ________ ________ ________

b Explain how you calculated 10 less.

____

5 **a** Write the number that is **100 less** than 2038. ______________

b Explain how you worked this out.

____

Hands and Feet

DATE:

You will need: a partner, a sheet of paper, a pencil

1 **a** Trace around your foot. Then, carefully trace around your hand, with your fingers together. Cut out the outlines.

b Use the foot and hand outlines to measure the objects in the table.

c Record your results.

d Select an item of your own to measure for the last box.

Item	How many hands?	How many feet?
Length of table 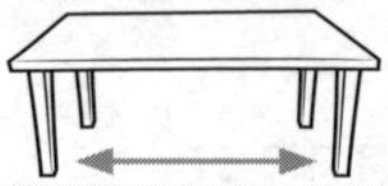		
Height of table 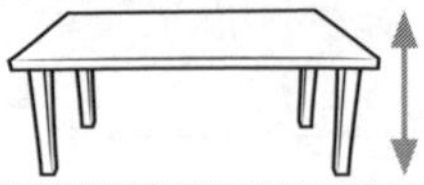		
Width of doorway 		
Height of doorway 		
Width of window 		
Width of display board 		
Width of bookshelf 		
Length of teacher's desk 		
Height of teacher's desk 		
My choice:		

2 **a** Compare your measurements with another pair's measurements.

b Were your results the same? ______________________

c How do you think that happened? ______________________

____

My Hand in Centimetres

DATE:

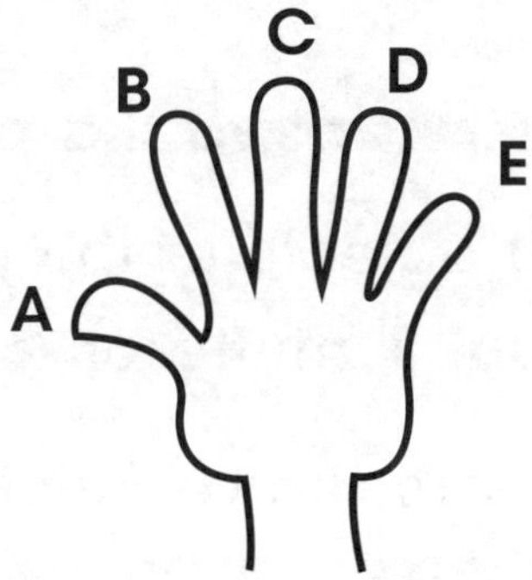

You will need: a ruler, a sheet of paper

1 Trace around your hand on a sheet of paper.

2 Rule a line on your hand to measure the following in **centimetres**:

a The **length** of your hand (from the bottom of your hand to the tip of your middle finger [C]).

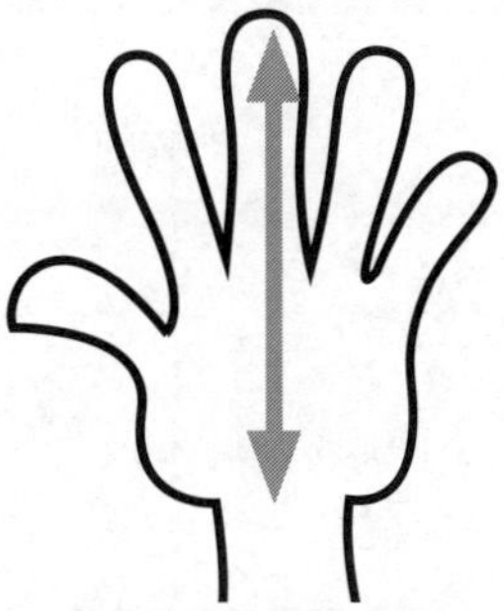

____

b The **width** of your hand (from one side of your palm to the other).

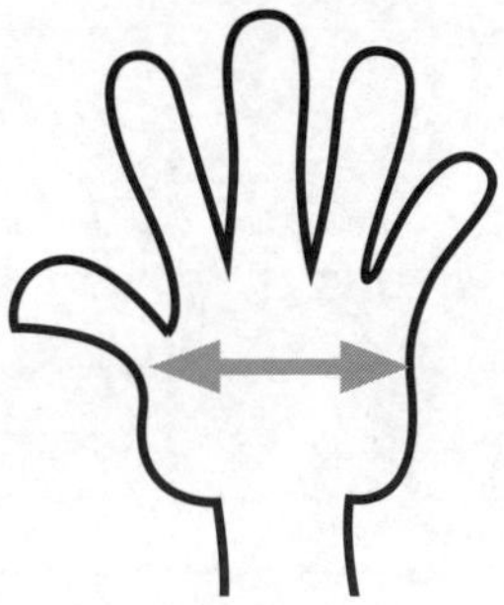

____

c **Your hand span** (from the tip of your thumb [A] to the tip of your little finger [E]).

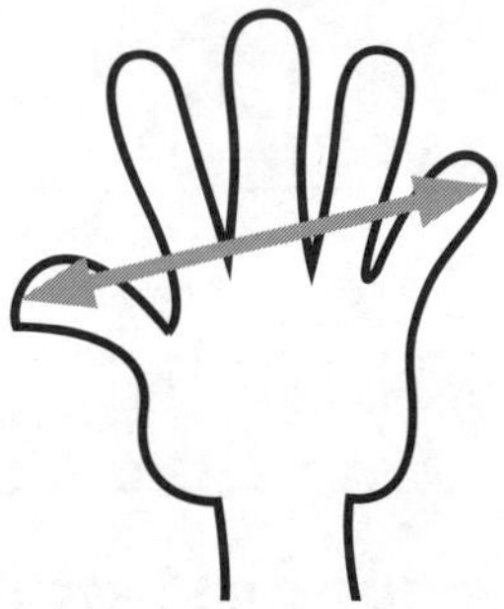

____

d The **length** of each of your fingers and your thumb.

A = __________ B = __________ C = __________ D = __________ E = __________

- Which is the **longest**? ______________ How long is it? ______________
- Which is the **shortest**? ______________ How long is it? ______________

How Long?

DATE:

You will need: a partner, 1-metre ruler or 1-metre length of string

Find objects in the classroom with a length **shorter than** 1 metre, **about** 1 metre or **longer than** 1 metre.

1 Name and sketch each item in the table below.

Shorter than 1 metre	About 1 metre	Longer than 1 metre

2 What was the **longest** object you measured? ______________________

3 What was the **shortest** object you measured? ______________________

4 Which category was the **easiest** to find? ____________________ Why?

____

5 Which category was the **hardest** to find? ____________________ Why?

____

Length (TRB pp. 34–37)
Using units of measurement Measure, order and compare objects using familiar metric units of length, area, mass and capacity
(VCMMG140)

Unit 4

DATE:

STUDENT ASSESSMENT

You will need: rulers

1 Name the two units of measurement we have used in this unit.

______________________ ______________________

2 Explain why it is important that everyone measures with the same units, rather than using the length of their hands or feet.

____

3 Measure the length of each line.

a ______________________ length: ________

b ______________________ length: ________

c ______________________ length: ________

d ______________________ length: ________

e ______________________ length: ________

f ______________________ length: ________

4 Order the lengths from Question 3 from **shortest** to **longest**.

____

5 Draw a line matching each measuring device to its name.

trundle wheel

30 cm ruler

1-metre ruler

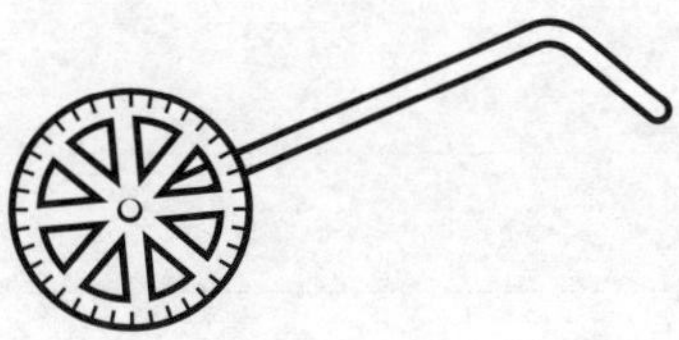

tape measure

Adding 10s

DATE:

1 Complete the table. The first line has been done.

Number	+ 10	+ 20	+ 50	+ 100
36	46	56	86	136
25				
53				
72				
Challenge! **243**				
Challenge! **357**				

2 Use the number lines to help you solve the addition problems.

Example

36 + 20 = 56

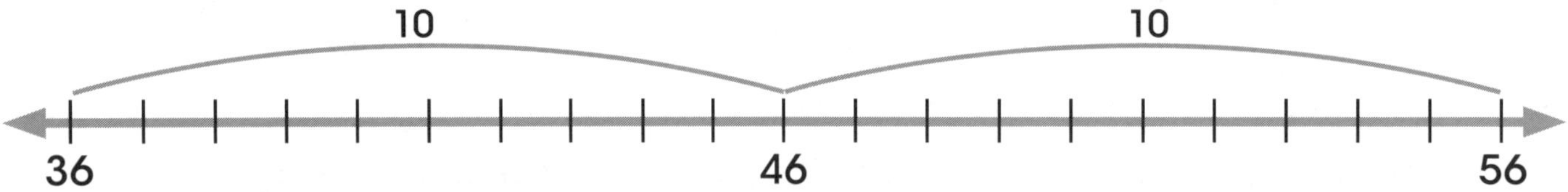

a 43 + 30 = ______________

b 39 + 30 = ______________

c 75 + 40 = ______________

d 62 + 20 = ______________

Build to 10

DATE:

1 Draw a line matching the numbers that are 10s facts.
One has been done.

5	**4**	**6**	**0**	**1**	**9**	**2**	**7**	**3**	**8**	**10**
8	**10**	**4**	**9**	**2**	**0**	**5**	**1**	**7**	**6**	**3**

2 How many more would be needed to build to the next 10?
One has been done.

a 46  50 — 4

b 37  40 — ______

c 62  70 — ______

d 79  80 — ______

e 22  30 — ______

f 85  90 — ______

3 Use the **build-to-10 strategy** to help you solve these addition problems.
One has been done.

Problem	Build to 10	Answer
57 + 6 =	57 + 3 + 3	63
74 + 9 =		
38 + 5 =		
73 + 9 =		
44 + 7 =		
69 + 3 =		
135 + 7 =		
268 + 5 =		

Challenge!

There were 27 ducks in the pond.
Then 8 more ducks came along.
How many ducks are there now?

Don't forget to use the **build-to-10 strategy**!

Doubles and Near Doubles

DATE:

You will need: a stopwatch or a clock with a second hand

1 Answer the doubles facts. How many can you answer in 2 minutes? ______

6 + 6 =	7 + 7 =	4 + 4 =	9 + 9 =	2 + 2 =
12 + 12 =	20 + 20 =	3 + 3 =	8 + 8 =	14 + 14 =
1 + 1 =	5 + 5 =	13 + 13 =	11 + 11 =	10 + 10 =
19 + 19 =	17 + 17 =	18 + 18 =	15 + 15 =	16 + 16 =

2 Write a **doubles fact** and a **near doubles fact** for each number. Don't forget to write the answer as well! One has been done.

Number	Doubles fact	Near doubles fact
6	6 + 6 = 12	6 + 7 = 13
9		
10		
15		
8		
16		
Challenge! 25		
Challenge! 40		

3 Draw a line matching the **near doubles fact** to the **doubles fact** that will help you solve it. Then solve each fact.

7 + 8 =

9 + 9 =

20 + 21 =

6 + 7 =

9 + 8 =

7 + 7 =

11 + 12 =

6 + 6 =

12 + 13 =

20 + 20 =

12 + 12 =

11 + 11 =

Unit 5

STUDENT ASSESSMENT

DATE:

1 Complete these wheels.

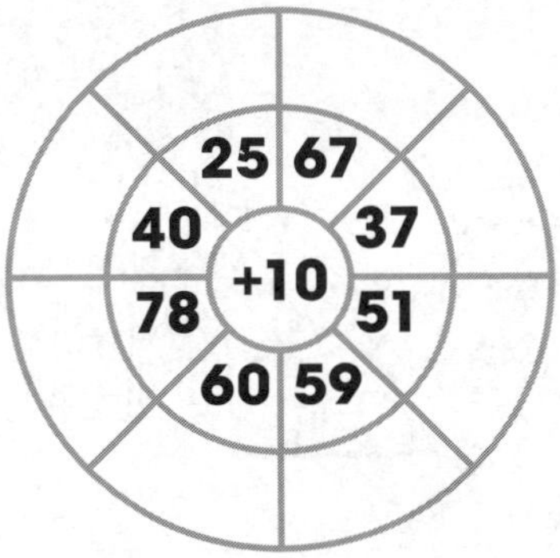

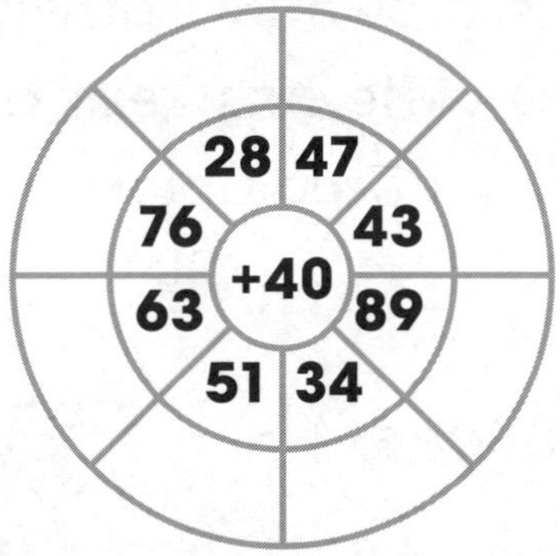

2 Use the number line to solve this addition problem.

51 + 47 =

3 Solve these addition problems and explain how you used the **build-to-10 strategy**.

45 + 6 =

67 + 8 =

4 Solve these near doubles facts. Write the doubles fact you used to help you.

Near doubles fact	Doubles fact that helped
15 + 16 =	
11 + 12 =	
25 + 26 =	

5 Solve the following. Write the strategy you used.

26 + 7 = Strategy: ____________

59 + 20 = Strategy: ____________

46 + 30 = Strategy: ____________

72 + 34 = Strategy: ____________

Addition in the Garden

DATE:

You will need: MAB blocks or **BLM 7 'MAB'**

Waleed has grown lots of vegetables in his garden. He has kept a record of how many he has picked.

tomatoes 48
corn 37
beans 65
carrots 16
potatoes 29
zucchinis 7

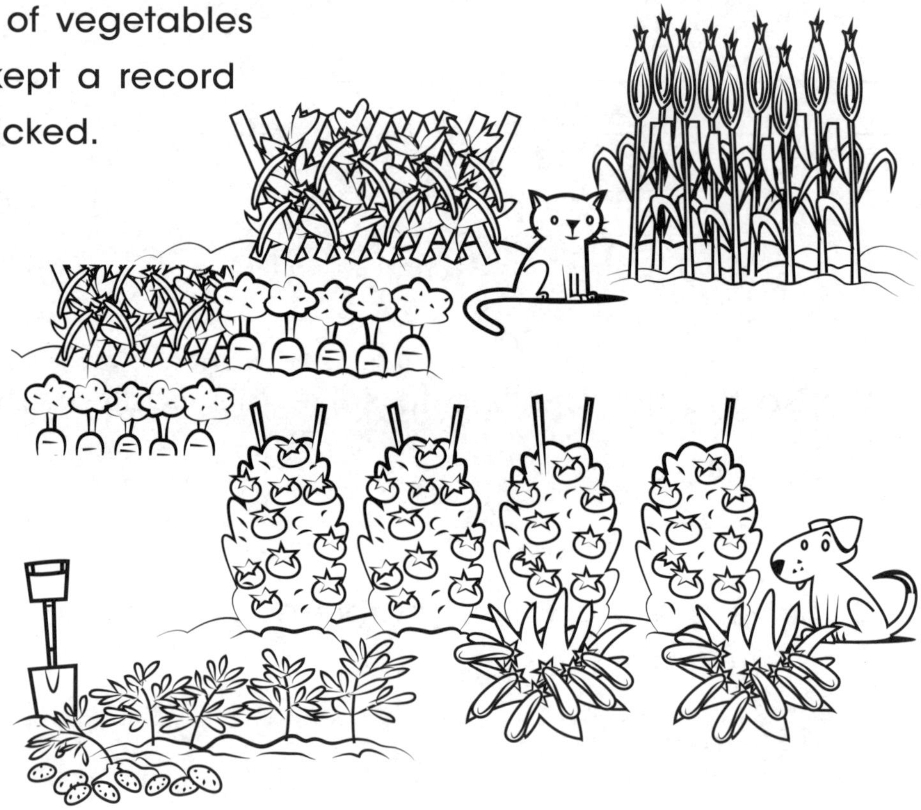

How many vegetables did Waleed pick in the following combinations?
Write the addition problem and use MAB blocks to work out the answer.

carrots and potatoes	potatoes and beans	potatoes and corn
beans and tomatoes	tomatoes and corn	tomatoes, carrots and beans
corn and zucchinis	zucchinis and carrots	beans, zucchinis and corn

Challenge!

How many vegetables did Waleed pick altogether? ____________

Addition (TRB pp. 42–45)
Number and place value Recognise and explain the connection between addition and subtraction (VCMNA132)

Which Number?

DATE:

Solve each problem with one of the numbers in the table.
Record the matching letter to find the answer to the joke!

What is a prehistoric monster called when it sleeps?

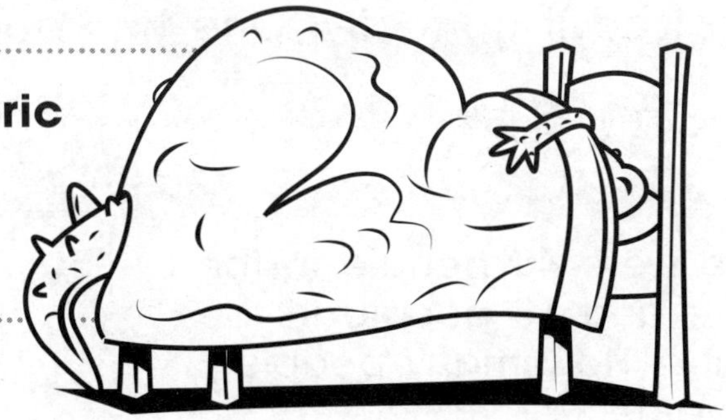

28 + 35

____

36 + 22	45 + 19	32 + 28	48 + 15	19 + 24	17 + 26	39 + 20
____	____	____	____	____	____	____

25 + 74	29 + 14	46 + 13	37 + 23	56 + 27	18 + 31	54 + 45
- ____	____	____	____	____	- ____	____

40 + 20	57 + 26	58 + 35
____	____	____

A 63	**B** 53	**C** 42	**D** 7	**E** 83	**F** 61	**G** 80	**H** 71	**I** 35	**J** 19	**K** 28	**L** 31	**M** 9
N 43	**O** 59	**P** 4	**Q** 11	**R** 60	**S** 99	**T** 58	**U** 49	**V** 1	**W** 3	**X** 93	**Y** 64	**Z** 5

Estimating Problems

DATE:

Estimate the answer to each addition problem by rounding the numbers up or down. Solve each problem and use your estimate to decide if you solved it correctly. One has been done.

Problem	Estimate by rounding	Solve it!	Was I accurate? ✓ or ✗
There were 48 people in the house and 13 people in the garden. How many people altogether?	50 + 10 = 60	48 + 13 = **61**	✓
I saw 17 birds sitting on the fence and 29 birds flying over the house. How many birds were there?			
Peter spent $47 at the shops on Saturday and $53 on Sunday. How much did he spend on the weekend?			
Ling rode her bike for 37 minutes and then jumped with her skipping rope for 29 minutes. How many minutes did she do exercise for?			
There were 57 apples on the tree and 38 that had fallen onto the ground. How many apples altogether?			
Challenge!		59 + 26	
Challenge!	40 + 80 =		

Addition (TRB pp. 42–45)
Number and place value Recognise and explain the connection between addition and subtraction (VCMNA132)

Unit 6

DATE:

STUDENT ASSESSMENT

1 Solve these addition problems.

a
48
\+ 31

b
25
\+ 57

c
63
\+ 18

d
37
14
\+ 23

2 Check the following problems and fix any mistakes.

a
27
\+ 36
53

b
52
\+ 39
81

c
48
\+ 29
67

d
34
\+ 17
41

3 What things do you need to remember when solving or checking addition problems?

____

____

4 Estimate the answer to each addition problem. Then draw a line to what you think the correct answer might be. One has been done.

Problem	My estimate	Answer
48 + 23	50 + 20 = 70	77
26 + 72		63
49 + 28		71
68 + 22		98
24 + 39		90

Writing Directions

DATE:

1 Write directions to help someone find their way around the classroom. Choose a Start and Finish location.

Write the directions, step by step, to get from Start to Finish.

Ask a partner to begin at Start and follow your directions.

Start location ____________ **Finish location** ____________

Directions

Did your partner get to the Finish location? Yes / No

2 Now choose a different Start and Finish location and repeat the task.

Start location ____________ **Finish location** ____________

Directions

Did your partner get to the Finish location? Yes / No

Map of My Bedroom

DATE:

1 a Circle the items that you have in your bedroom.

b Draw some other items that you have in your bedroom.

2 a On another sheet of paper, draw a map of your bedroom. Don't forget to show where the door is. Make sure you include all the items you circled and the items you drew.

b Label the items on your map.

c Explain what you found difficult when drawing the map of your bedroom.

Using a Map

DATE:

Use the map to answer these questions.

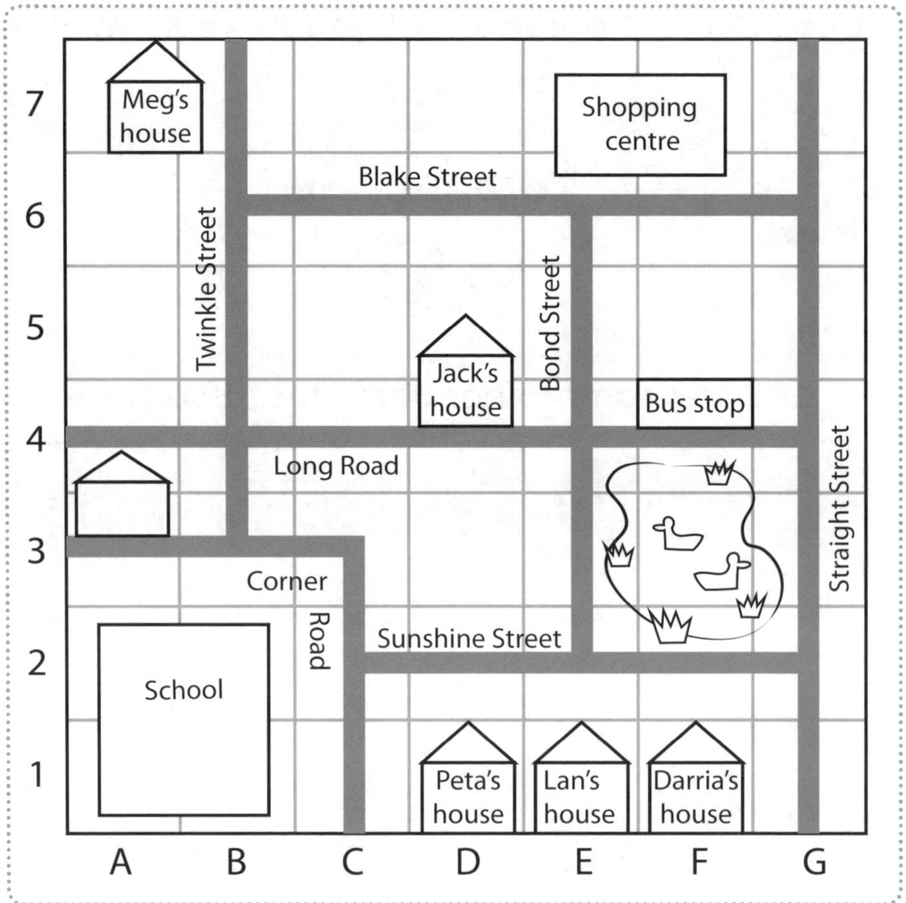

1 Who lives on Long Road? ____________

2 What street is the school on? ____________

3 Who are the three students who live next to each other? ____________

____

4 What are the coordinates for the duck pond? ____________

5 Who lives across the road from the duck pond? ____________

6 Tim lives across the road from the school. Write his name on his house.

7 Draw a large tree in the block next to Jack's house.

8 What coordinates would the children go to if they wanted to catch a bus? ____________

9 There is a big car park stretching from C7 to D7. Draw it in the correct place.

10 Draw the path Meg would take to get to Lan's house.

11 Draw the path you could take if you went from school to the shops.

12 Draw a milk bar at the corner of Straight Street and Blake Street.

13 Draw a playground that stretches from A5 to A6.

Location (TRB pp. 46–49)
Location and transformation Create and interpret simple grid maps to show position and pathways (VCMMG143)

DATE:

STUDENT ASSESSMENT

1 Write the directions you would give to someone who needed to find the office from your classroom.

____

____

2 Think of your bathroom at home. On another sheet of paper, draw a map of your bathroom.

3 Look at the map.

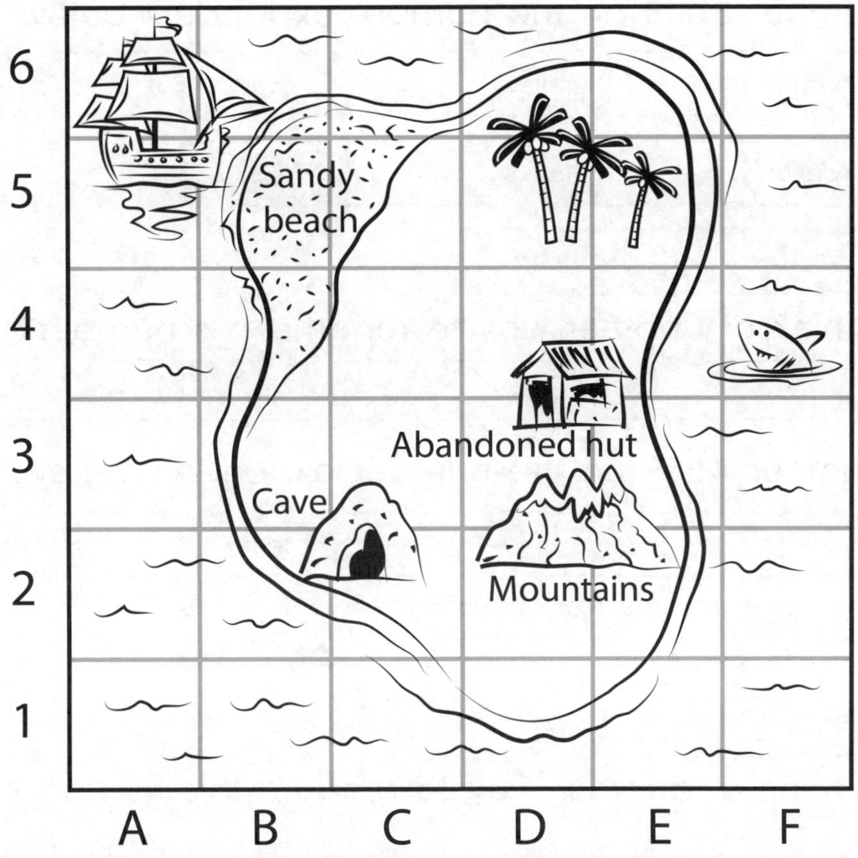

a What are the coordinates of the shark? __________

b What coordinates would be the best place on the island for the pirates to land? __________

c What is next to the cave? __________

d What is between the coconut trees and the mountains? __________

e The treasure is buried at C4. Mark the spot with a **✕**.

Using Number Expanders

DATE:

1 Show the number 5 163 in different ways on the number expanders below.

a

	thousands		hundreds		tens		ones

b

	thousands		tens		ones

c

	thousands		ones

d Can you think of another way to rename this number?

____

2 Show the number 3 638 on the number expanders below.

a

	thousands		hundreds		tens		ones

b

	thousands		tens		ones

c

	hundreds		ones

d Can you think of another way to rename this number?

____

3 Show the number 4 559 on the number expanders below.

a

	thousands		hundreds		tens		ones

b

	thousands		tens		ones

c

	ones

d Can you think of another way to rename this number?

____

Challenge!

Rename these numbers, using no thousands and only 2 tens in each.

a 5 613 ______________________________

b 3 638 ______________________________

c 4 559 ______________________________

Cars for Sale!

DATE:

1 Order the car prices from **largest** to **smallest**.

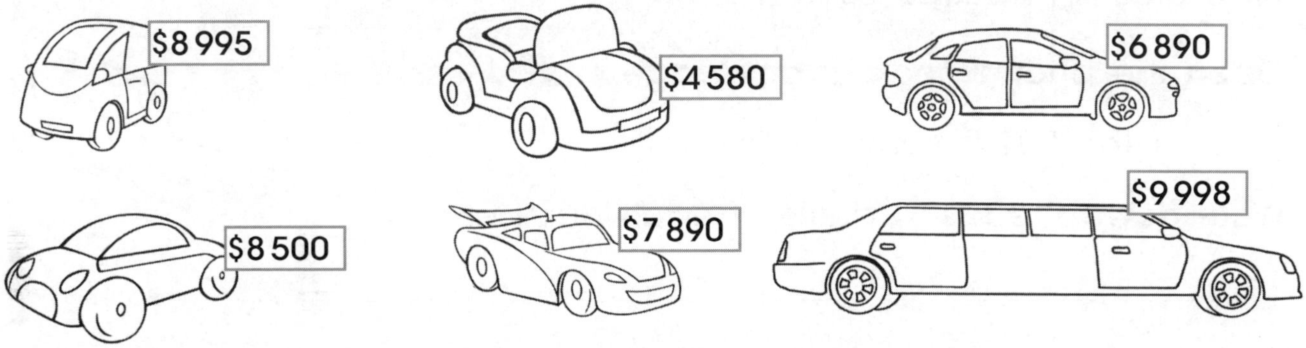

____

2 Write the following numbers in words.

a 6 890 ______________________________

b 9 998 ______________________________

c 4 580 ______________________________

3 Write the value of **8** in each of these numbers.

a 6 890 ____________ **b** 9 998 ____________

c 4 580 ____________ **d** 7 890 ____________

e 8 995 ____________ **f** 8 500 ____________

4 Partition these numbers. One has been done.

a 6 890 = 6 000 + 800 + 90

b 9 998 = ________ + ________ + ________ + ________

c 4 580 = ________ + ________ + ________ + ________

d 7 890 = ________ + ________ + ________ + ________

e 8 995 = ________ + ________ + ________ + ________

f 8 500 = ________ + ________ + ________ + ________

Number Hunt

DATE:

You will need: a phone book (or telephone directory on the internet), MAB blocks or **BLM 7 'MAB'**

1 Open the phone book at any page.

Look at the first 9 numbers listed.

Write down the last 4 digits of each number.

__________ __________ __________

__________ __________ __________

__________ __________ __________

2 a Use **BLM 7 'MAB'** or MAB blocks to model each number from Question 1.

b Record the numbers from **smallest** to **largest** on the lines in the first column. Then, add or subtract the number in the next column. Write the answer on the line provided. Use MAB blocks to solve problems if needed.

smallest	__________	*add 30*	__________
	__________	*add 400*	__________
	__________	*add 2000*	__________
	__________	*subtract 20*	__________
	__________	*add 300*	__________
	__________	*add 90*	__________
	__________	*add 700*	__________
	__________	*add 6000*	__________
largest	__________	*subtract 800*	__________

DATE:

STUDENT ASSESSMENT

1 Use the number expander to rename 6930. Then, follow the instructions below.

	thousands		hundreds		tens		ones

a with no tens ______

b with no hundreds ______

c with no hundreds or thousands ______

d with only ones ______

2 Write the value of **9** in each of these numbers.

a 6937 ______ b 2895 ______

c 1964 ______ d 2549 ______

e 3190 ______ f 9060 ______

3 Partition these numbers.

a 6937 = ______ + ______ + ______ + ______

b 2895 = ______ + ______ + ______ + ______

c 1964 = ______ + ______ + ______ + ______

d 2549 = ______ + ______ + ______ + ______

e 3190 = ______ + ______ + ______ + ______

f 9060 = ______ + ______ + ______ + ______

4 Complete the following. You may use MAB blocks to model the numbers.

a 6937 *add 30* ______ b 2549 *add 3* ______

c 2895 *add 5000* ______ d 3190 *add 20* ______

e 1964 *subtract 800* ______ f 9060 *subtract 100* ______

Ancient Egyptian Hieroglyphs

DATE:

You will need: BLM 25 'Number Hieroglyphs'

1 Using the digits 2, 3, 5 and 6, create three different 4-digit numbers.

a ________ **b** ________ **c** ________

2 Refer to **BLM 25 'Number Hieroglyphs'** and write each of your numbers in hieroglyphs below.

a

b

c

3 Add 200 to your first number **(a)** and show your answer in hieroglyphs.

4 Subtract 50 from your second number **(b)** and show your answer in hieroglyphs.

5 Add 3 300 to your third number **(c)** and show your answer in hieroglyphs.

Place-value Path

DATE:

You will need: MAB blocks or **BLM 7 'MAB'**

Fill in the missing numbers and operations to complete the path. Model the numbers with MAB as you work. Note: operations need to be written in place-value terms.

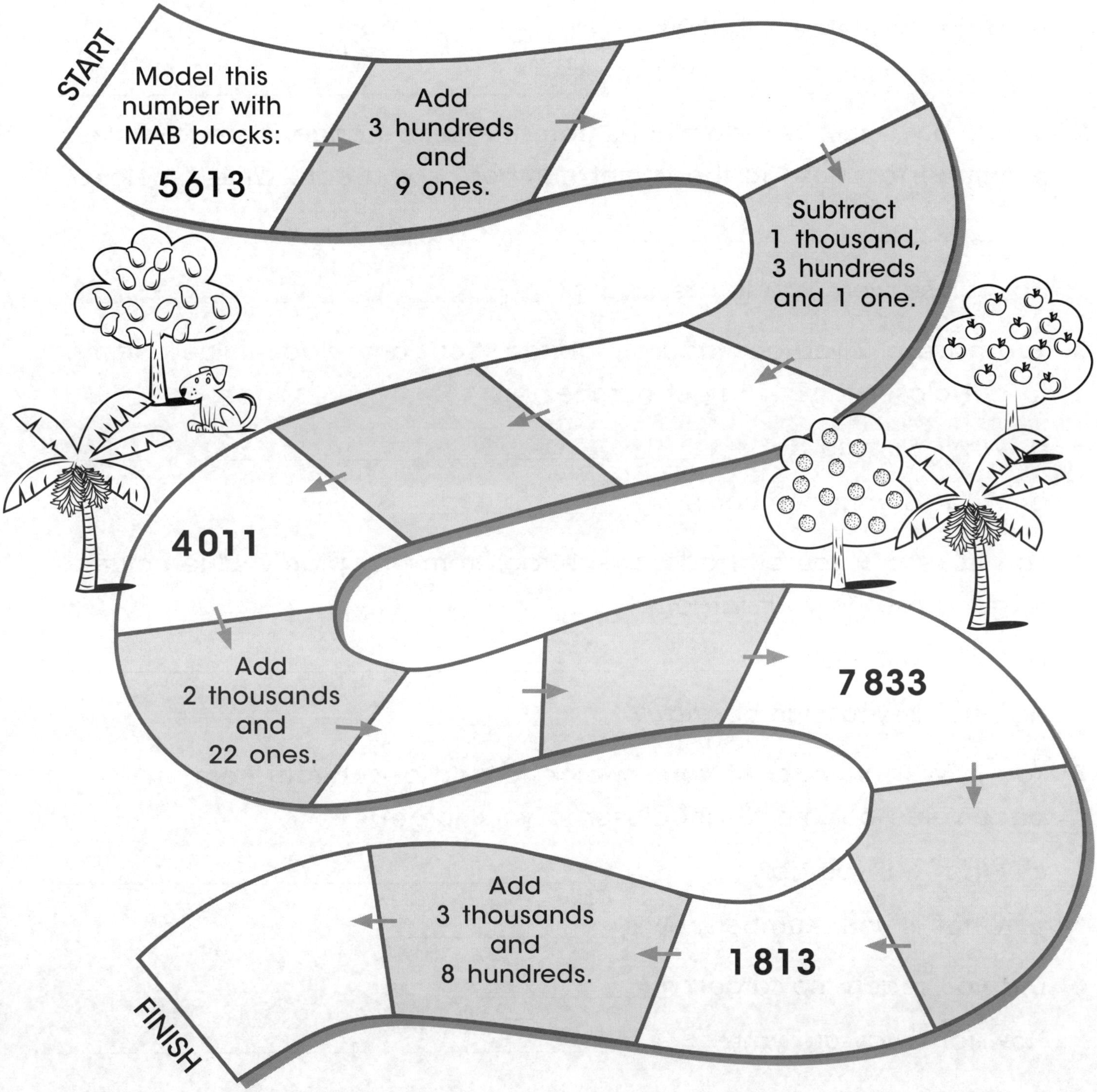

Reach the Target

DATE:

You will need: a set of dominoes or **BLM 27 'Dominoes'** (cut up)

1 The dominoes show your target number.

a What number is shown on the dominoes? ______________

b Write the number in words.

____

2 Randomly select two dominoes from the set. Arrange them to make a number as close to the target number as you can. Write it below.

____

How close were you to the target? ______________________________

3 You have a **2** you can add or subtract from any place-value column to get closer to your target number.

a What will you do? ______________________________

b What is your number now? ______________

4 You have a **5** you can add or subtract from any place-value column to get closer to your target.

a What will you do? ______________________________

b What is your number now? ______________

5 You have a **number of your choice** to add or subtract from any pace-value column to get closer to your target.

a What will you do? ______________________________

b What is your number now? ______________

6 Did you reach the target? ______________

How far away are you? ______________________________

Unit 9

DATE:

STUDENT ASSESSMENT

1 Using the code in the box, write the number that is represented.

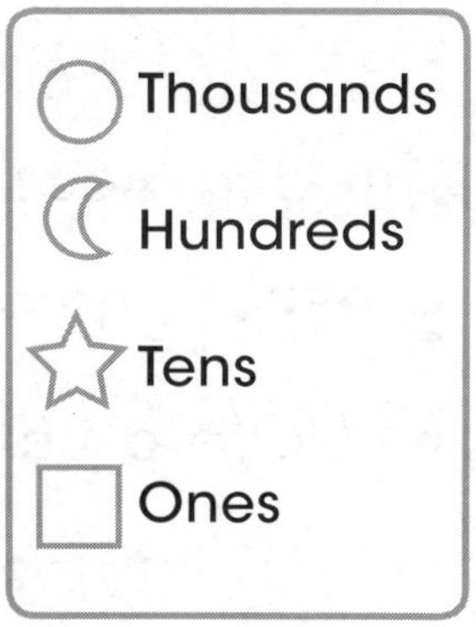

a 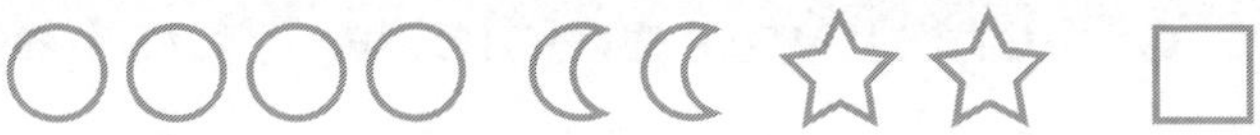

____

b 

____

2 Write the following numbers using the code.

a 2179 ______________________________

b 3476 ______________________________

c 5098 ______________________________

3 Complete the following. You may use MAB blocks to model the numbers.

a Add 3 hundreds and 2 tens to 2179. __________

b Subtract 4 hundreds, 3 tens and 4 ones from 3476. __________

c Add 3 ones to 5098. __________

4 Complete the sentences.

The target number is 2341. If you started with the number 4586 ...

- You would need to add/subtract __________ thousands to reach the target.
- You would need to add/subtract __________ hundreds to reach the target.
- You would need to add/subtract __________ tens to reach the target.
- You would need to add/subtract __________ ones to reach the target.

Which Is Heaviest?

DATE:

You will need: a beam balance, classroom items shown below

1 Compare the two items in each group.

- **a** Circle the item (or group of items) you predict will be **heaviest**.
- **b** Compare the items on the beam balance.
- **c** Place a tick ✔ next to your prediction if you were correct.

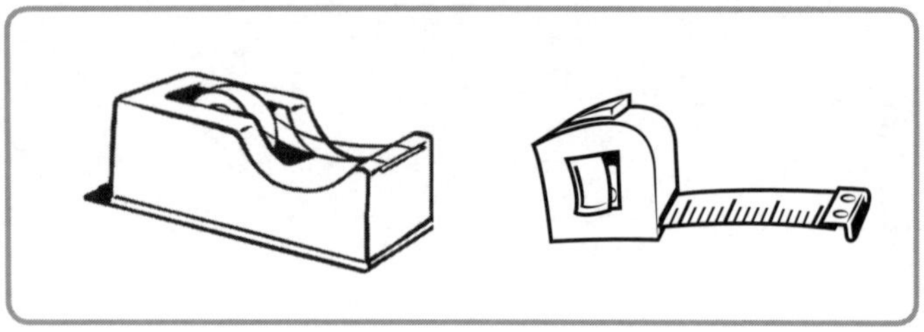

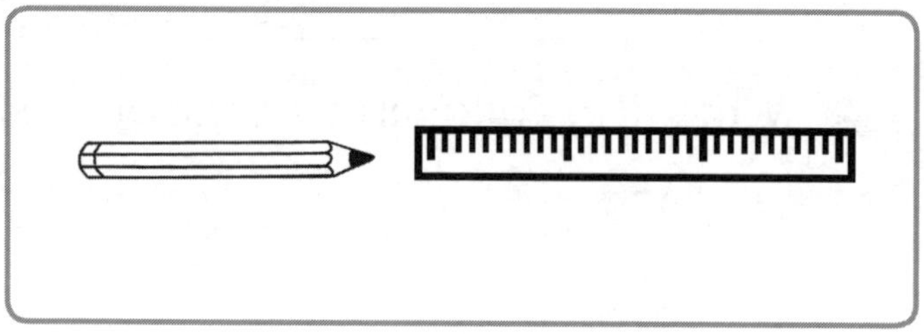

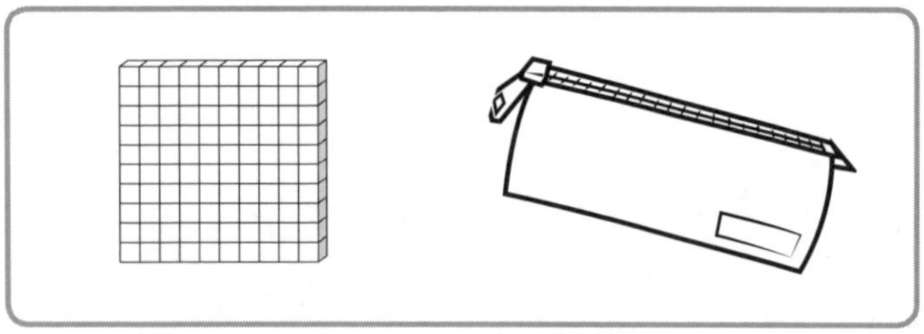

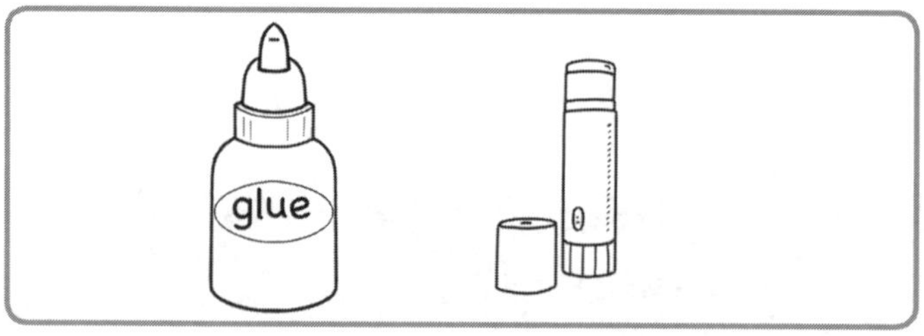

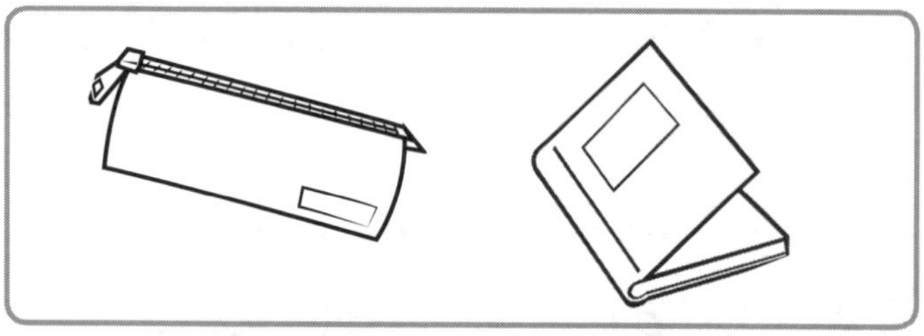

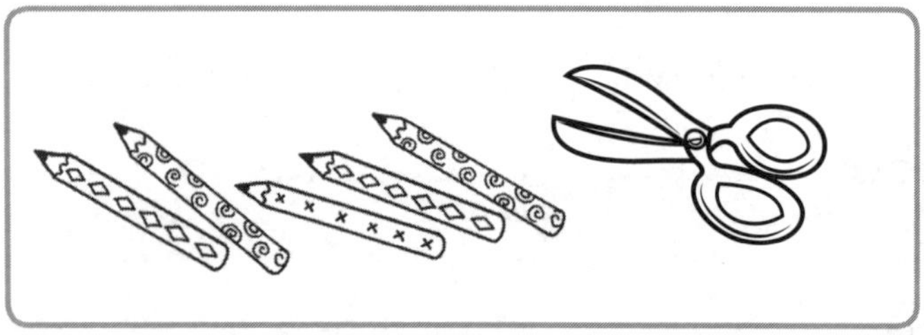

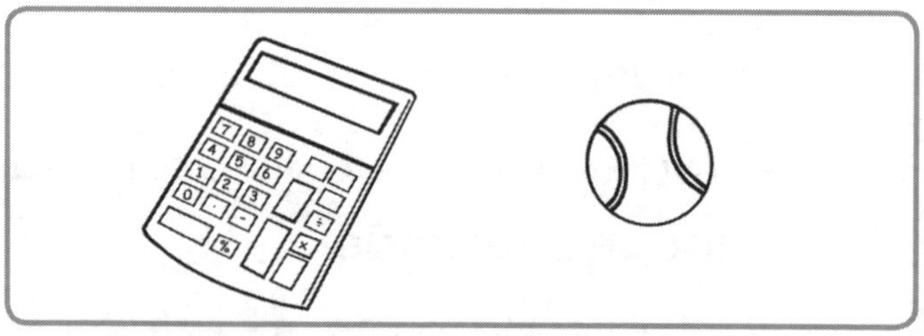

2 Find 3 things in the classroom that are **lighter than** a pencil.

__________________ __________________ __________________

3 Find 3 things in the classroom that are **heavier than** a stapler.

__________________ __________________ __________________

Mass (TRB pp. 58–61)
Using units of measurement Measure, order and compare objects using familiar metric units of length, area, mass and capacity
(VCMMG140)

Weighing in Grams

DATE:

You will need: a beam balance, a set of gram weights, items below

1 Fill in the following table. You need to:

- predict the order from lightest (1) to heaviest (6) by **looking**.
- predict the order from lightest (1) to heaviest (6) by **feeling**.
- **weigh** each item to find the actual mass (in grams).
- number each item in order from lightest (1) to heaviest (6).

Item	Predict order by looking (1–6)	Predict order by feeling (1–6)	Actual weight (g)	Actual order (1–6)
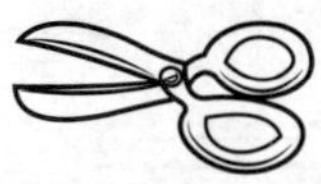				

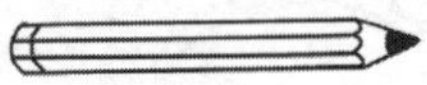				
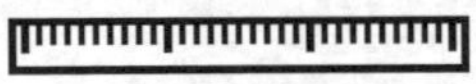				

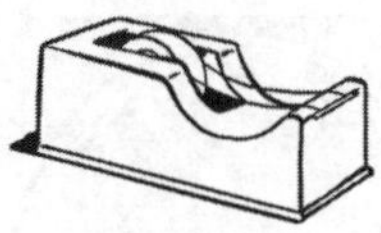				

2 Were your predictions from looking or feeling more accurate? Why?

____

3 Were any of the items difficult to weigh accurately? Why?

____

Reading and Using Scales

DATE:

You will need: kitchen scales or a beam balance and weights, classroom items

1 Write the mass shown on each set of scales.

2 Find an item (or group of items) in the classroom that matches the weight shown and draw this on the scales.

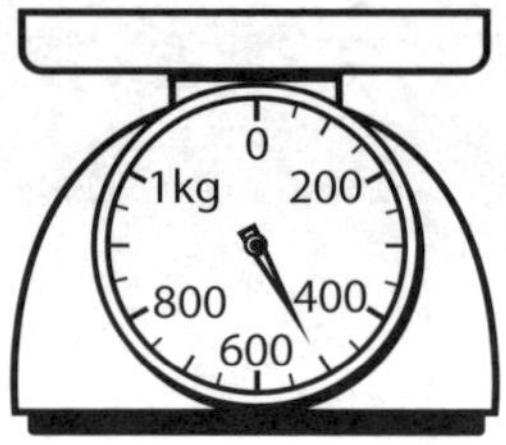

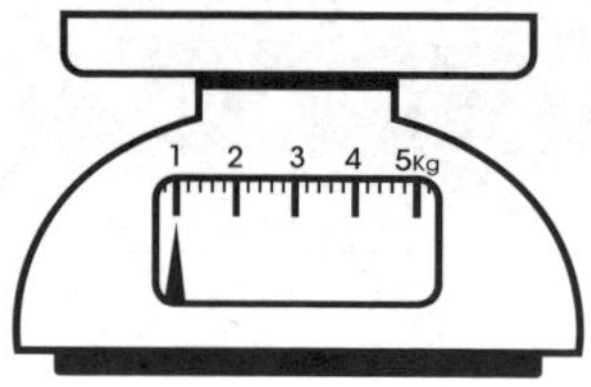

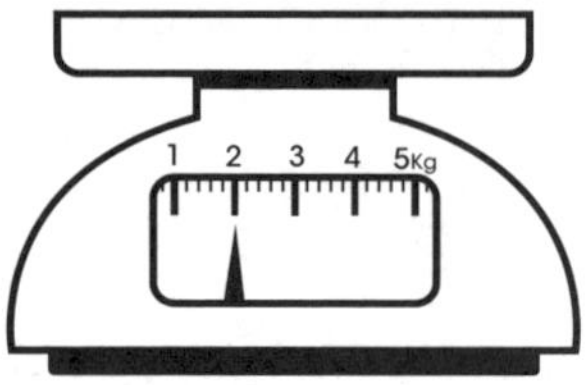

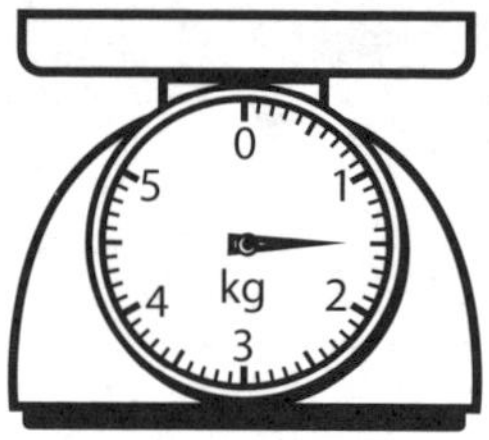

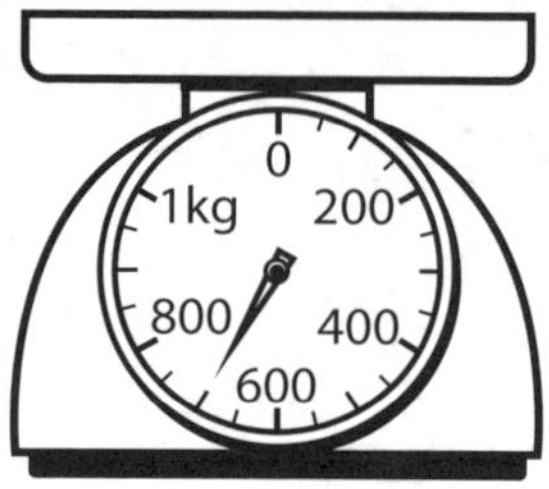

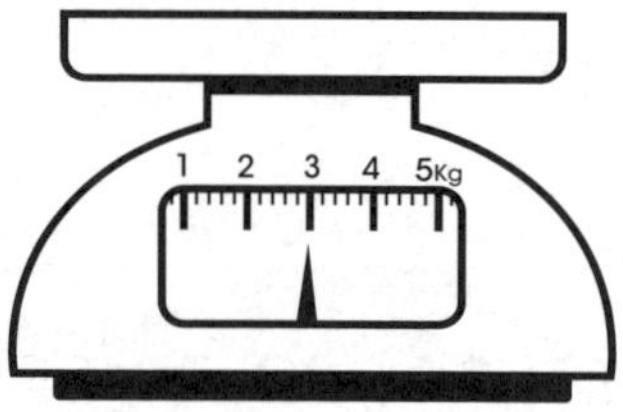

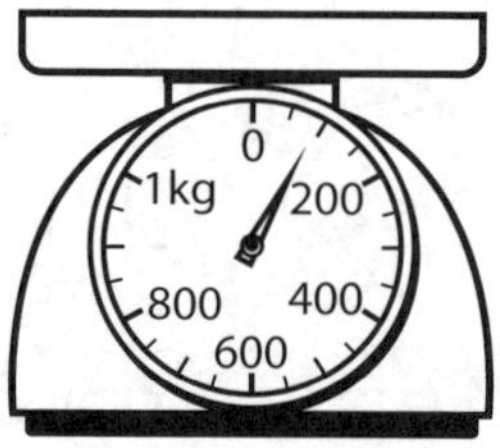

3 Were there any weights that were difficult to find? Why?

DATE:

STUDENT ASSESSMENT

1 Circle the **heaviest** item in each group.

a

b

c

d

2 Name 3 items that are measured in kilograms.

____________ ____________ ____________

3 Name 3 items that are measured in grams.

____________ ____________ ____________

4 Number the following items from **lightest** (1) to **heaviest** (5).

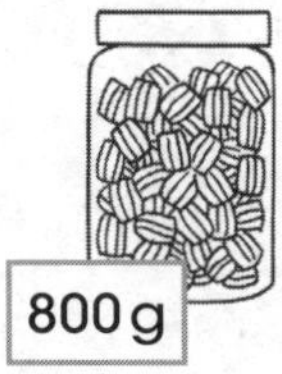

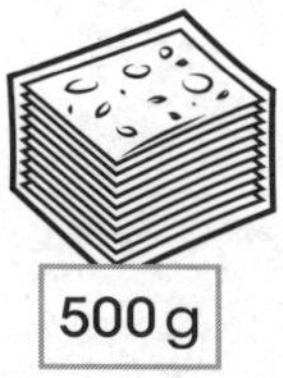

________ ________ ________ ________ ________

5 Write the mass shown on each set of scales.

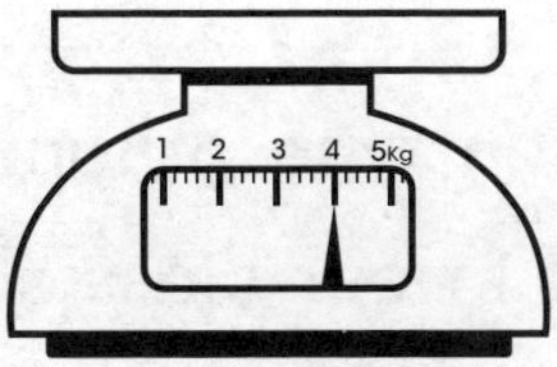

____________ ____________ ____________

Collecting Data

DATE:

1 **Research Question:** What season were most of our class members born in?

a Survey your classmates (or use a class birthday chart) to fill in the data collection table.

Category	Tally	Total
summer (Dec, Jan, Feb)		
autumn (Mar, Apr, May)		
winter (Jun, Jul, Aug)		
spring (Sep, Oct, Nov)		

b In which season were the **most** students born? ____________________

c In which season were the **least** students born? ____________________

2 **Research Question:** What kinds of furniture do we have in our classroom?

a Look around the classroom. Create 4 categories for the types of furniture in your classroom (e.g. chairs, tables, beanbags).

b Create a tally showing the number of each of these and write the total.

Category	Tally	Total

c What is the most common type of furniture? ____________________

d Complete this sentence: In our classroom, there are more ______________ than ________________.

Data representation and interpretation Identify questions or issues for categorical variables. Identify data sources and plan methods of data collection and recording (VCMSP148)
Collect data, organise into categories and create displays using lists, tables, picture graphs and simple column graphs, with and without the use of digital technologies (VCMSP149)

Creating a Picture Graph

DATE:

1 **Research Question:** How did Year 3 students get to school today?

Survey your classmates to fill in the data collection table. (You may add a mode of transport to suit your school.)

Mode of transport	Tally	Total
car 		
bus 		
train 		
pushbike 		
foot 		

2 On a blank sheet of A4 paper, create a **pictograph** to represent your data.

- **a** Write your graph title at the top of your page.
- **b** Draw your graph **axes**: vertical and horizontal. Make your graph as big as you can.
- **c** Decide on a symbol to use for each mode of transport. Make sure it is easy to draw.
- **d** Draw your symbols on your graph to represent your data.

A pictograph is a special kind of column graph which uses pictures instead of blocks to represent data. For example:

Favourite Fruit

Fruit

Interpreting Data

DATE:

Look at the graph below. Think about what it could represent.

Title: ____________________

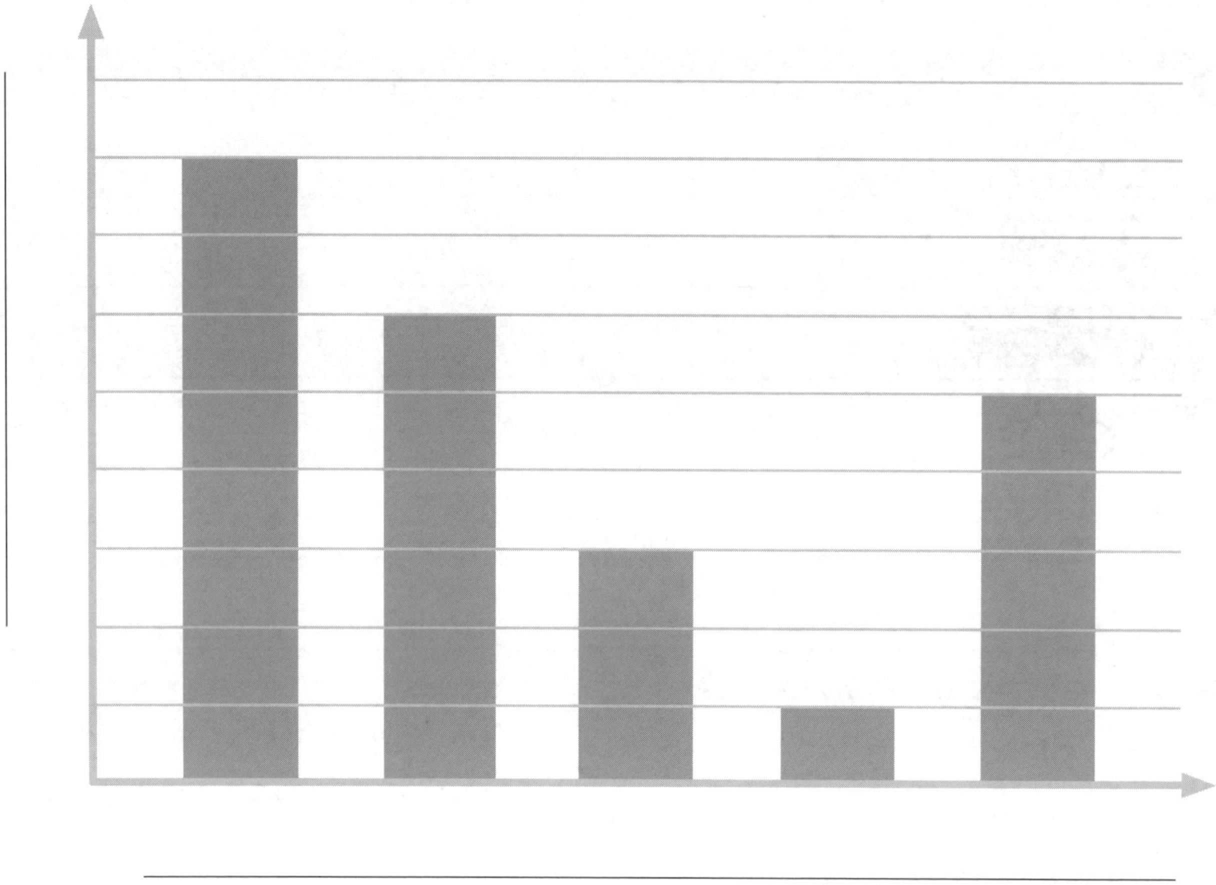

1 Give the graph a title.

2 Label both axes.

3 Label the categories on the horizontal axis.

4 Add numbers (scale) to the vertical axis.

5 Write 3 facts shown by the graph.

DATE:

STUDENT ASSESSMENT

Research Question: What is Year 3's favourite sport?

1 Fill in the tally in the data table.

Category	Tally	Total
basketball		10
football		9
netball		6
swimming		12
dancing		7

2 Create a column graph to represent the data.

Title: ____________________

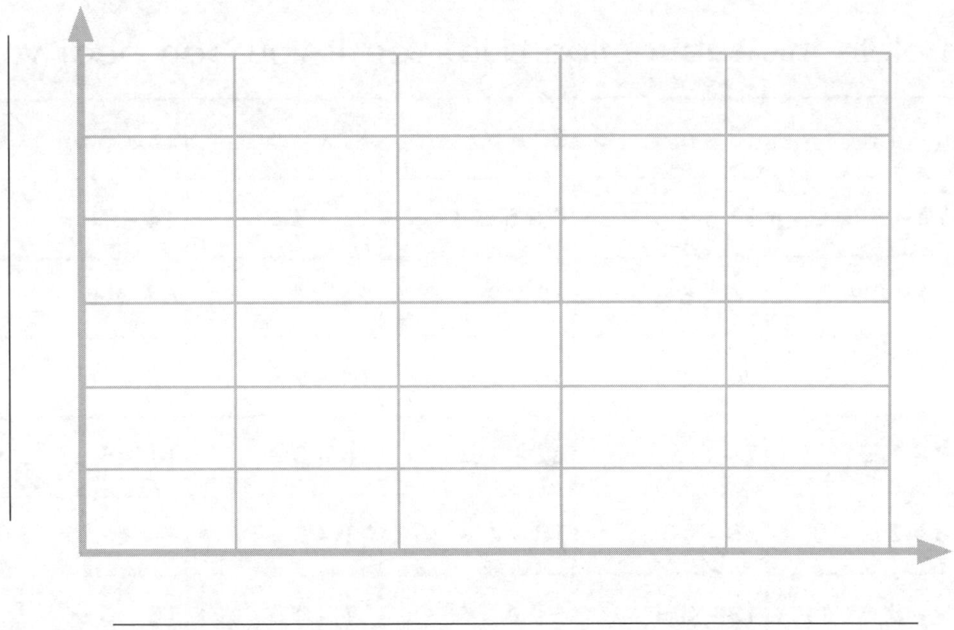

3 Write 3 facts shown by the graph.

____

____

____

Ready, Set, Subtract!

DATE:

You will need: a stopwatch or timer

1 Complete this subtraction grid as fast as you can! Use the timer.

12 - 3 =	7 - 4 =	19 - 6 =	12 - 10 =	8 - 4 =	24 - 6 =	25 - 10 =
6 - 4 =	10 - 8 =	17 - 4 =	11 - 4 =	22 - 7 =	19 - 4 =	13 - 5 =
7 - 3 =	15 - 3 =	11 - 6 =	19 - 3 =	12 - 6 =	28 - 7 =	21 - 5 =
10 - 5 =	13 - 8 =	29 - 6 =	15 - 4 =	35 - 5 =	22 - 4 =	6 - 4 =
13 - 7 =	24 - 6 =	13 - 9 =	29 - 8 =	8 - 4 =	13 - 5 =	16 - 8 =
9 - 8 =	11 - 7 =	8 - 6 =	14 - 4 =	26 - 9 =	17 - 4 =	13 - 4 =
25 - 8 =	9 - 6 =	13 - 7 =	27 - 5 =	14 - 5 =	35 - 3 =	8 - 4 =

I completed the grid in ________ minutes.

2 Now complete this subtraction grid. See if you can beat your last time!

14 - 6 =	6 - 4 =	13 - 6 =	20 - 5 =	17 - 3 =	13 - 5 =	18 - 5 =
24 - 7 =	19 - 8 =	11 - 9 =	8 - 5 =	11 - 3 =	14 - 10 =	5 - 3 =
23 - 2 =	16 - 4 =	14 - 5 =	12 - 9 =	24 - 7 =	6 - 4 =	13 - 5 =
9 - 6 =	17 - 4 =	26 - 8 =	30 - 10 =	14 - 9 =	22 - 6 =	17 - 9 =
13 - 6 =	8 - 1 =	7 - 5 =	22 - 9 =	16 - 3 =	11 - 5 =	28 - 2 =
7 - 2 =	6 - 3 =	15 - 10 =	18 - 4 =	11 - 8 =	7 - 4 =	9 - 1 =
34 - 6 =	9 - 5 =	13 - 3 =	6 - 6 =	37 - 4 =	15 - 6 =	5 - 1 =

I completed the grid in ________ minutes.

3 What strategies did you use to solve the subtraction problems?

____

____

Serial Subtraction

DATE:

You will need: BLM 7 'MAB', scissors, glue

1 Use the MAB blocks from **BLM 7 'MAB'** to complete number sentences in the tables. One number sentence has been done.

a

Number sentence	Model first number and cross off 5 blocks
38 - 5 = 33	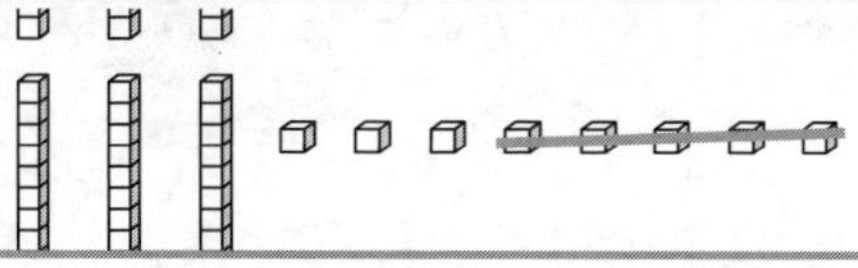
48 - 5 =	
58 - 5 =	
68 - 5 =	

b

Number sentence	Model first number and cross off 4 blocks
26 - 4 =	
36 - 4 =	
46 - 4 =	
56 - 4 =	

c What did you notice as you solved these equations?

____

2 Solve the following equation groups.

a 27 - 6 = ______ 37 - 6 = ______ 47 - 6 = ______ 57 - 6 = ______

b 19 - 3 = ______ 29 - 3 = ______ 39 - 3 = ______ 49 - 3 = ______

Orchard Farmer

DATE:

On Sunday, Mandi counted the number of pieces of fruit on each tree and then wrote the amount on the sign. Each day of the week she will pick a different amount of fruit. Mandi wants to know how much fruit will be left on the trees at the end of each day.

	Apples	Oranges	Mangoes	Bananas
Monday 10 pieces of fruit from each tree	87 - 10 = 77 77 apples left on the tree.			
Tuesday 20 pieces of fruit from each tree	77 - 20 = 57 57 apples left on the tree.			
Wednesday 30 pieces of fruit from each tree				
Thursday 11 pieces of fruit from each tree				
Friday 9 pieces of fruit from each tree				

Mental Strategies for Subtraction (TRB pp. 66–69)
Number and place value Recall addition facts for single-digit numbers and related subtraction facts to develop increasingly efficient mental strategies for computation
(VCMNA133)

Unit 12

DATE:

STUDENT ASSESSMENT

1 Draw lines matching the subtraction problems with their answers.

11 - 7   21 - 4   12 - 5   24 - 5   18 - 5   15 - 9

17   6   13   4   7   19

2 What strategies can you use to solve subtraction problems?

____

3 Complete the following number sentences.

a 25 - 4 = 21 and 35 - 4 = 31, so 45 - 4 = ____________.

b 38 - 5 = 33 and 48 - 5 = 43, so 58 - 5 = ____________.

4 Complete the table.

	- 10	**- 20**	**- 30**	**- 50**
67				
71				
94				

5 Use the number lines to help solve these subtraction problems.

a 75 - 41 =

b 68 - 29 =

At the Fair

DATE:

You will need: MAB blocks or **BLM 7 'MAB'**, extra paper for working out

1 Quin has $50 to spend. He can't decide what to spend his money on! He wants to know how much change he will have if he spends money on the following things.

Ferris wheel $50 - $24 = $26	hot dog	giant slide	jumping castle
bumper cars	balloon	ice-cream	ghost train

Write the subtraction problem for each and use MAB blocks or **BLM 7 'MAB'** to help you solve it. One has been done.

Challenge!

2 How much change would Quin have if he went on the giant slide, bought a balloon and bought an ice-cream? ______________________

3 What 2 things did Quin spend money on if he has change of $32?

____

Correction Time

DATE:

You will need: a calculator

1 Solve the following subtraction problems.

a $76 - 32$

b $54 - 27$

c $83 - 35$

2 Petra has completed her work. Check her work and write the correct answer.

a $47 - 29 = 22$

b $85 - 34 = 51$

c $78 - 34 = 43$

d $56 - 19 = 37$

e $36 - 12 = 22$

f $61 - 27 = 46$

g $82 - 44 = 42$

h $73 - 46 = 27$

i $64 - 9 = 45$

j $44 - 16 = 38$

Did Petra make any mistakes? ______________________

Why do you think she made those mistakes? ______________________

____

What things were you looking for as you corrected her work? __________

____

3 Now use a calculator to check if you worked out each answer in Question 2 correctly. Tick each problem you got correct.

4 Circle the correct answer to these subtraction problems.

a $83 - 24$

43 59 61

b $62 - 46$

16 24 26

c $43 - 17$

34 36 26

Match the Estimation

DATE:

Write an equation for each problem. Draw a line to the equation you would use to find an estimate. Use it to check your answer.

There were 48 people travelling on a train. 19 people got off when it stopped at the station. How many people were left on the train?

Franco is saving for a new bike which costs $83. He has $49 in his money box. How much more money does he need?

Carl ran around the oval in 51 seconds and Jenna ran around it in 26 seconds. How many seconds faster was Jenna than Carl?

Ling has 92 stamps in her collection and Manni has 33 stamps in his collection. How many more does Ling have than Manni?

Estimations

60 - 40 = 20

80 - 50 = 30

90 - 30 = 60

50 - 30 = 20

50 - 20 = 30

50 - 40 = 10

70 - 30 = 40

Scott collected 63 shells when he went to the beach. There was a hole in his bag and he lost 38 of them. How many did he have left?

There were 68 fish in an aquarium. 34 of them were moved into another tank. How many fish were left?

There were 47 apples growing on the tree. 21 fell onto the ground during a storm. How many were left on the tree?

Maya has 37 basketball cards. She needs 50 cards to have the complete set. How many more does she need?

Unit 13

DATE:

STUDENT ASSESSMENT

1 Solve the following subtraction problems.

a $\begin{array}{r} 75 \\ -\ 21 \\ \hline \end{array}$ **b** $\begin{array}{r} 64 \\ -\ 38 \\ \hline \end{array}$ **c** $\begin{array}{r} 81 \\ -\ 26 \\ \hline \end{array}$ **d** $\begin{array}{r} 52 \\ -\ 46 \\ \hline \end{array}$

2 Check the following problems and correct any mistakes.

a $\begin{array}{r} 82 \\ -\ 56 \\ \hline 34 \end{array}$ **b** $\begin{array}{r} 72 \\ -\ 19 \\ \hline 63 \end{array}$ **c** $\begin{array}{r} 64 \\ -\ 37 \\ \hline 33 \end{array}$ **d** $\begin{array}{r} 52 \\ -\ 28 \\ \hline 34 \end{array}$

3 What things do you need to remember when solving or checking subtraction problems? ______________________________

____

4 Estimate the answers to these subtraction problems. Then choose the correct answer from the box below.

a 48 - 31 = Estimate: ________ Correct answer: ________

b 78 - 34 = Estimate: ________ Correct answer: ________

c 39 - 11 = Estimate: ________ Correct answer: ________

d 62 - 28 = Estimate: ________ Correct answer: ________

e 71 - 49 = Estimate: ________ Correct answer: ________

f 86 - 37 = Estimate: ________ Correct answer: ________

17	28	22	44	34	49

5 How can estimating help you solve subtraction problems?

____

Fact Families

DATE:

1 Write the facts that can be made using these numbers.

a 7 5 12

____ + ____ = ____ ____ + ____ = ____

____ - ____ = ____ ____ - ____ = ____

b 9 6 15

____ + ____ = ____ ____ + ____ = ____

____ - ____ = ____ ____ - ____ = ____

c 8 5 13

____ + ____ = ____ ____ + ____ = ____

____ - ____ = ____ ____ - ____ = ____

d 5 6 11

____ + ____ = ____ ____ + ____ = ____

____ - ____ = ____ ____ - ____ = ____

e 10 8 18

____ + ____ = ____ ____ + ____ = ____

____ - ____ = ____ ____ - ____ = ____

2 Write the missing problem to complete each family.

a 13 + 4 = 17
17 - 13 = 4
4 + 13 = 17
____

b 15 - 9 = 6
9 + 6 = 15
15 - 6 = 9
____

c 18 - 7 = 11
7 + 11 = 18
11 + 7 = 18
____

3 Solve each problem. Colour each family of fish a different colour.

7 + 4 =	8 + 5 =	6 + 3 =	11 - 4 =
9 - 3 =	13 - 8 =	4 + 7 =	5 + 8 =
3 + 6 =	11 - 7 =	13 - 5 =	9 - 6 =

Connections Between Addition and Subtraction (TRB pp. 74–77)
Number and place value Recognise and explain the connection between addition and subtraction (VCMNA132)

Recall addition facts for single-digit numbers and related subtraction facts to develop increasingly efficient mental strategies for computation (VCMNA133)

Related Addition and Subtraction

DATE:

1 **a** Slip down the slide by solving the addition and subtraction problems!

b Solve a **related problem** for each problem to check if your answer is correct. One has been done.

2 Draw a string matching each child to a balloon that will help them check the answer to their problem. One has been done.

Balancing Equations

DATE:

1 Draw a set of scales underneath each equation to show if it is balanced or uneven.

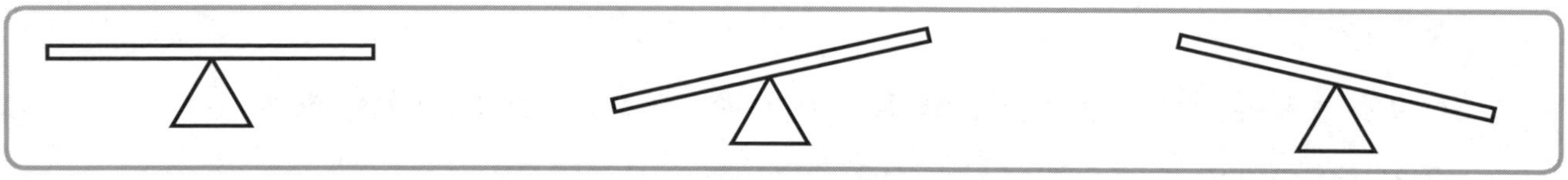

a 9 + 5 and 16 − 2 **b** 11 − 3 and 5 + 4 **c** 24 + 5 and 19 + 10

d 18 − 5 and 13 + 9 **e** 24 + 4 and 32 − 4 **f** 7 + 2 + 4 and 15 − 6

2 Fill in the missing numbers to balance the scales.

a 14 + 5 = 11 + ______

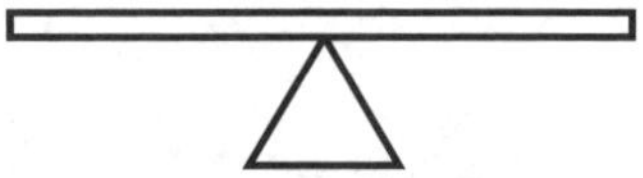

b 10 + 8 = 20 − ______

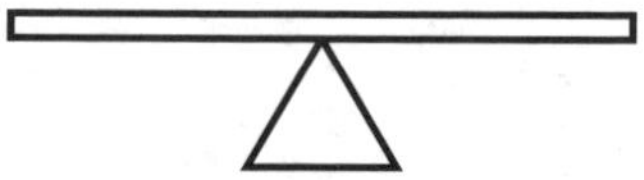

c 15 − 4 = 7 + ______

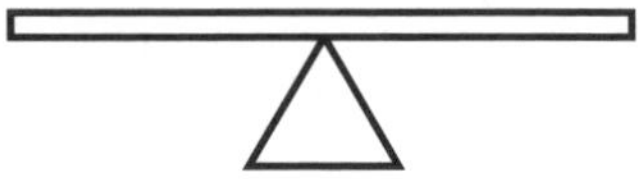

d 12 + 6 = ______ + 13

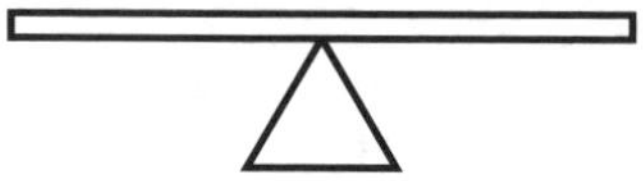

e 15 − ______ = 17 − 6

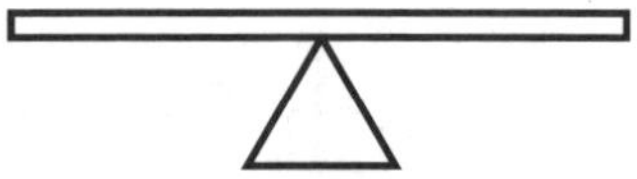

f 36 + 3 = 5 + ______

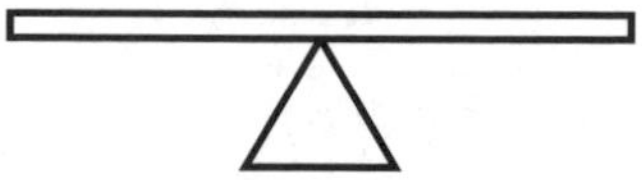

g 60 − 22 = 30 + ______

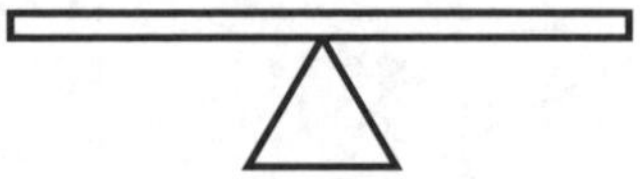

h 34 + 7 = 25 + ______ + 1

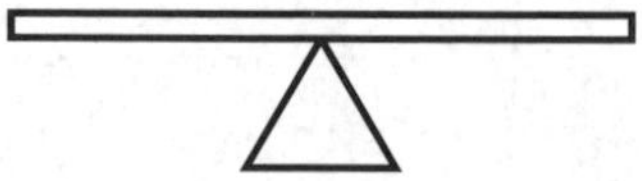

3 Write your own problems to balance the scales.

a 15 + 7 = ______ + ______ + ______

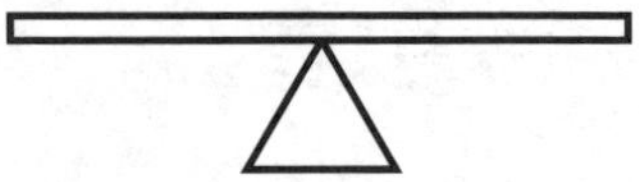

b 24 − 8 = ______ − ______

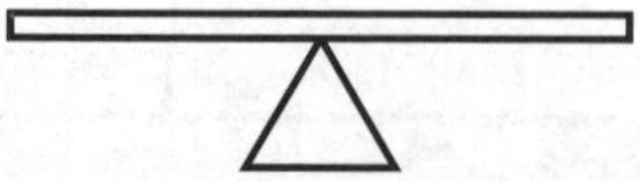

DATE:

STUDENT ASSESSMENT

1 Write fact families using the following numbers.

a 9, 4 and 13

____ + ____ = ____

____ + ____ = ____

____ - ____ = ____

____ - ____ = ____

b 6, 8 and 14

____ + ____ = ____

____ + ____ = ____

____ - ____ = ____

____ - ____ = ____

c 7, 4 and 11

____ + ____ = ____

____ + ____ = ____

____ - ____ = ____

____ - ____ = ____

2 Solve these problems. Check your answer by forming and solving a related problem.

a
$$\begin{array}{r} 46 \\ +\ 28 \\ \hline \end{array}$$

b
$$\begin{array}{r} 72 \\ -\ 35 \\ \hline \end{array}$$

3 What is the missing number in each problem?

a
$$\begin{array}{r} 67 \\ +\quad\ \ \\ \hline 82 \end{array}$$

b
$$\begin{array}{r} 59 \\ -\quad\ \ \\ \hline 35 \end{array}$$

c
$$\begin{array}{r} 28 \\ +\quad\ \ \\ \hline 47 \end{array}$$

d
$$\begin{array}{r} 82 \\ -\quad\ \ \\ \hline 63 \end{array}$$

4 How did you work out the missing number in each problem?

____

5 Balance these equations.

a 15 + 6 = 11 + ____

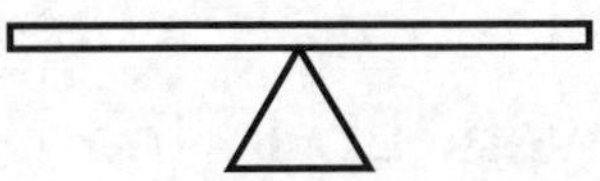

b 23 - ____ = 14 + 3 + 2

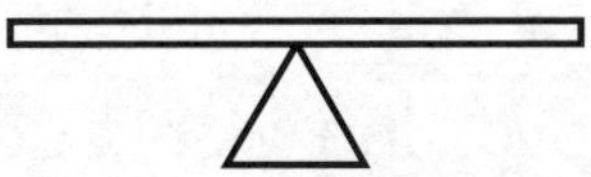

c 25 + 36 = 63 - ____

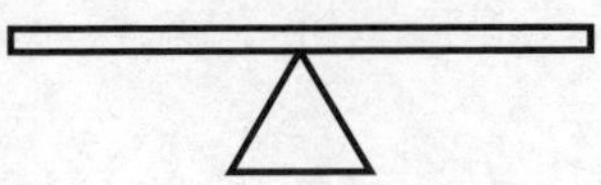

d 49 - ____ = 13 + 4 + 2

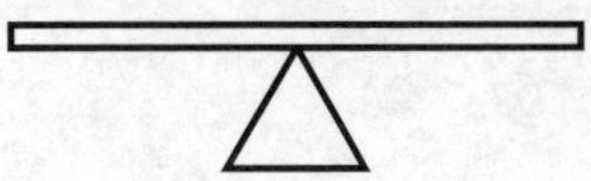

Addition Problems

DATE:

1 Solve these addition problems. Circle the problems with regrouping.

Draw a ★ next to the problems with no regrouping.

a 357 + 269

b 526 + 172

c 148 + 629

d 418 + 289

e 323 + 498

f 287 + 451

g 521 + 336

h 761 + 198

2 Colour the balloons that have a total of 635.

429 + 206

357 + 276

263 + 372

378 + 257

296 + 339

276 + 359

481 + 154

129 + 510

392 + 241

501 + 134

457 + 178

316 + 319

230 + 405

3 Solve each of these problems using an addition equation.

a There were 116 hot air balloons flying on Saturday and 128 on Sunday. How many balloons were in the sky over the weekend?	**b** 125 people went in hot air balloons in March, 235 in April and 142 in May. How many people went up in the balloons altogether?

Subtraction Problems

DATE:

What is hiding in the picture below?

Solve the subtraction problems. Use the code to colour the picture.

If the answer is 418 colour the space blue.
If the answer is 123 colour the space light green.
If the answer is 325 colour the space orange.
If the answer is 247 colour the space dark green.
If the answer is 132 colour the space grey.

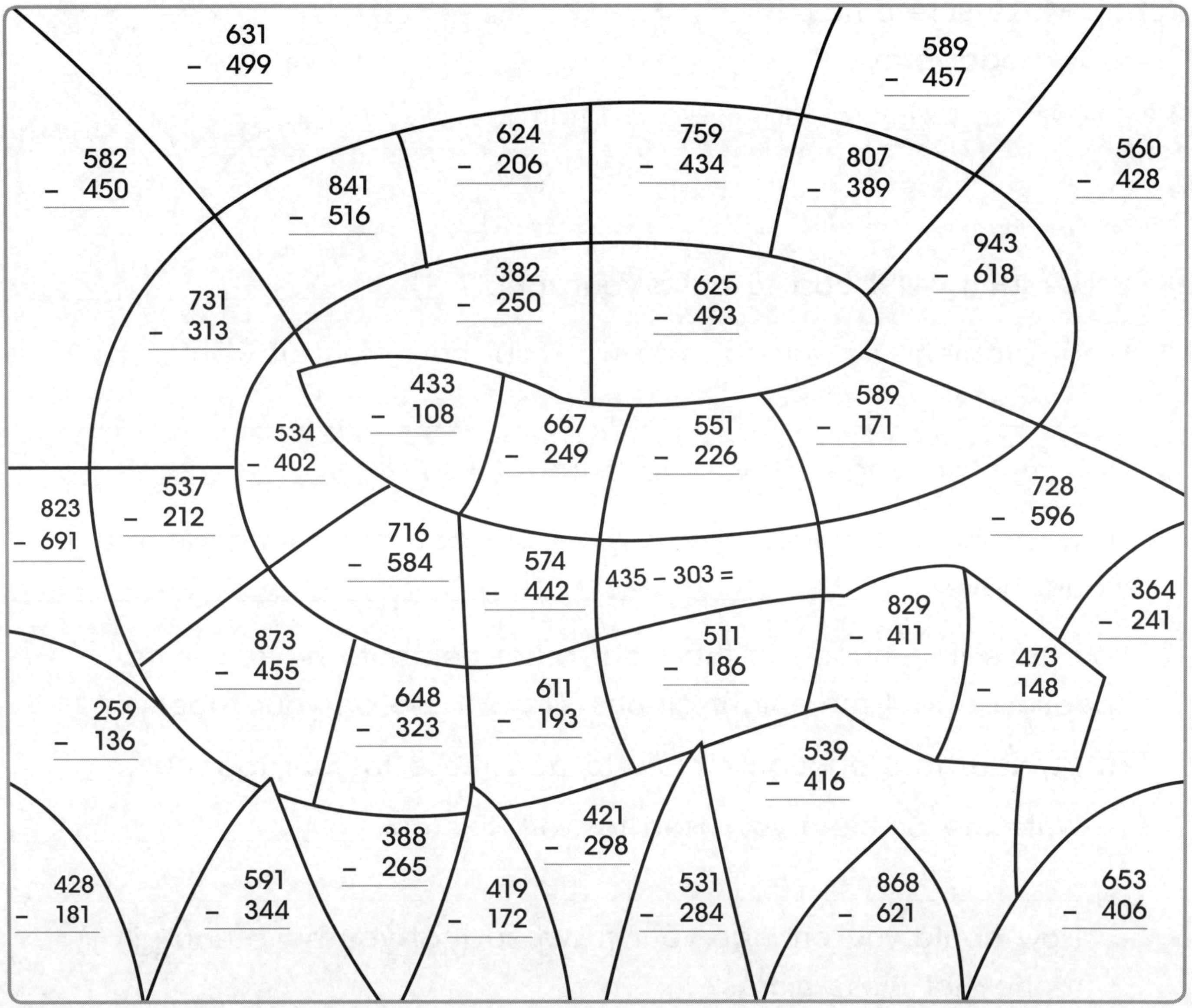

Build a Robot

DATE:

You will need:

BLM 41 'Robot Parts'

You have been given $1 000 to build a robot! Choose your robot parts from **BLM 41 'Robot Parts'**, then draw a picture of your robot. Make sure it has a body, head, arms and legs!

1 How much will it cost to build your robot? ______________________

2 What problem did you solve to work out how much it would cost?

____

3 How much change will you have from $1 000? ______________________

4 What problem did you solve to work out how much change you would have? ______________________

5 You have just found out that your robot needs to have 3 extra speakers and 4 more light globes. Draw these on your robot.

a Do you have enough money to add these to your robot? __________

b Write the problem you used to work this out. ______________________

____

c How could you change your robot so that you have enough money for these extras? ______________________

DATE:

STUDENT ASSESSMENT

1 Solve these addition problems.

a 468 + 231 = ______

b 673 + 252 = ______

c 285 + 167 = ______

c 576 + 348 = ______

2 Solve these subtraction problems.

a 546 - 123 = ______

b 472 - 345 = ______

c 735 - 468 = ______

d 626 - 159 = ______

3 Solve these problems.

a There were 125 pencils in a box and 175 in a pencil case. How many pencils were there altogether?	**b** Sam had $235 dollars. He spent $123 on a new pair of shoes. How much money does he have left?

How did you know whether to use addition or subtraction to solve these problems? ______________________________

____

4 There are 128 children in Year 3 and 225 children in Year 4. They are all going on an excursion to the museum. There are enough seats on the buses for 430 children. How many spare seats will there be? ______________

How did you work out the answer? ______________________________

____

In One Minute ...

DATE:

You will need: a partner, a stopwatch

1. Estimate how long you think it will take to do each task. Shade the box with a coloured pencil to show your estimate.
2. Complete each task. Have your partner do the timing.
3. Write the actual time in the appropriate box. For example: if it takes 40 seconds write 40 seconds in the 'Less than 1 minute' box.
4. Now add one task of your own and repeat.

Task	Less than 1 minute	About 1 minute	More than 1 minute
Clap 60 times.			
Untie and tie your shoelace.			
Do 10 star-jumps.			
Sing the national anthem twice.			
Say the alphabet 3 times.			
Say your 2-times tables 4 times.			
Write "My name is ______" 10 times.			

Challenge!

Write down a task you think you can do exactly 100 times in 1 minute. Use the stopwatch to check. ______________________________

____

Minutes on a Clock

DATE:

1 These analogue clocks show times between 4 and 5 o'clock. Write the matching digital time.

[ ][ ]:[ ][ ]   [ ][ ]:[ ][ ]   [ ][ ]:[ ][ ]   [ ][ ]:[ ][ ]   [ ][ ]:[ ][ ]

2 Add the minute hand to show the times on these clocks.

4:05   4:30   4:17   4:50   4:59

3 Of all the times shown in Questions 1 and 2:

a Which is the **earliest**? ______________

b Which is the **latest**? ______________

c List all of the times that are **between** 4:15 and 4:45.

____

4 Each of the clocks shown is within the hour of 4:00. Look carefully at the hour hand on each of the clocks in Questions 1 and 2.

a Are they all in the same position? Yes / No

b Why do you think the position of the hour hand changes?

____

Analogue Time

DATE:

1 Draw the missing hands on each clock. Make sure you position the hour hand accurately!

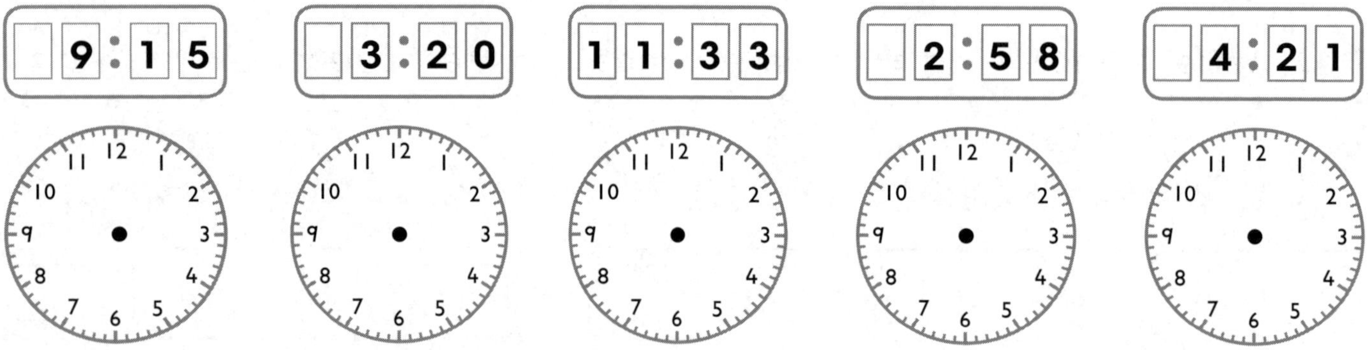

2 Write the correct time under each clock:

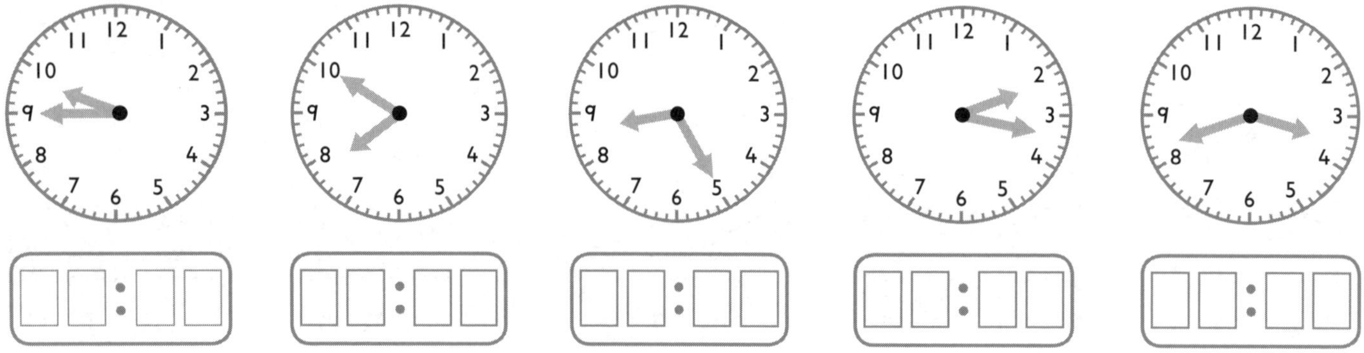

3 Oh no, my minute hand is missing!

a Work out the approximate time on each clock using only the hour hand.

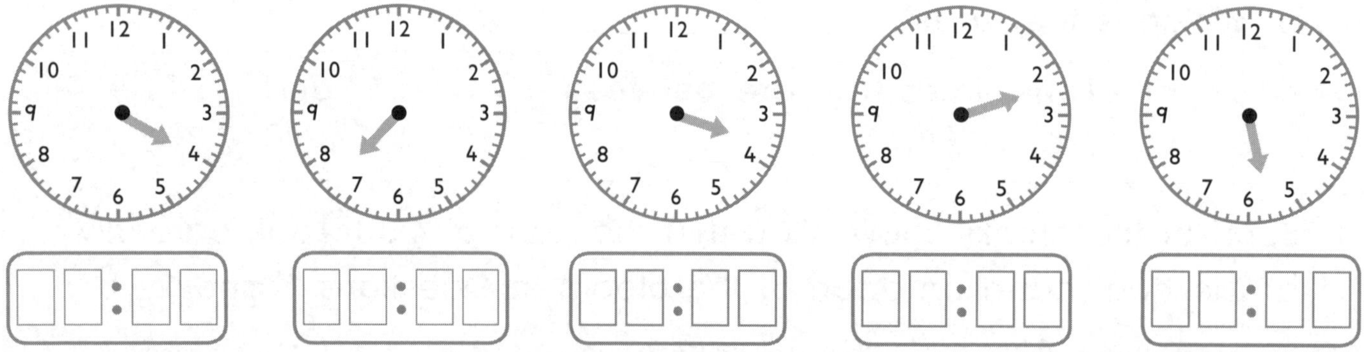

b Do you think this is more or less accurate than using both the minute hand and hour hand? Why? ______________________________

____

Time (TRB pp. 82–85)
Using units of measurement Tell time to the minute and investigate the relationship between units of time (VCMMG141)

DATE:

STUDENT ASSESSMENT

1 **a** How many seconds are there in 1 minute? ____________

b How many minutes are there in 1 hour? ____________

2 Write the following times in seconds into the correct box.

34 seconds	55 seconds	75 seconds
19 seconds	63 seconds	2 seconds
81 seconds	40 seconds	100 seconds

Less than 1 minute	About a minute	More than 1 minute

3 Write the times shown on these clocks.

☐☐:☐☐ ☐☐:☐☐ ☐☐:☐☐ ☐☐:☐☐ ☐☐:☐☐

4 Draw the missing hands on each clock:

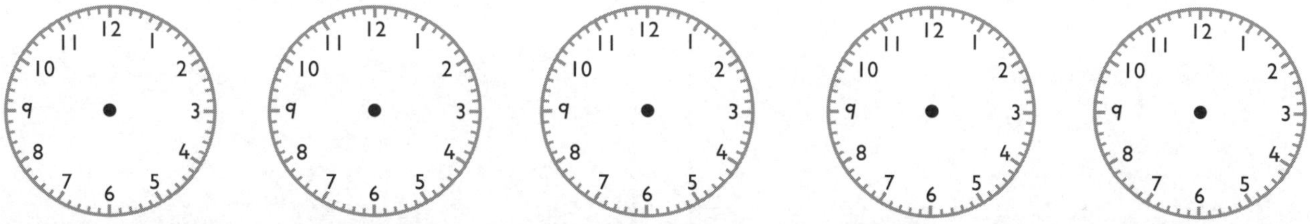

10:15 10:35 9:20 9:50 7:02

5 Think of two activities we measure in:

a seconds: ____________ ____________

b minutes: ____________ ____________

c hours: ____________ ____________

Drawing Lines

DATE:

1 Draw these pictures. Make sure you use the correct lines!

a Draw a house using 5 horizontal lines, 6 vertical lines, 2 diagonal lines and 1 curved line.

b Draw a cat using 3 curved lines, 6 diagonal lines, 1 vertical line and 1 horizontal line.

c Draw a car using 4 curved lines, 4 vertical lines, 3 horizontal lines and 2 diagonal lines.

d Draw a bike using 4 curved lines, 6 diagonal lines, 4 vertical lines and 3 horizontal lines.

2 Have a partner see if they can find all the lines in your pictures.

Were they able to find them all? Yes / No

Where Are the Angles?

DATE:

1 Search for the angles in the picture. Colour them in.

2 How many angles did you find? ____________________

3 What was the most common type of angle? Draw it in the box.

Angle Time

DATE:

You will need: a clock with movable hands or a clock made from **BLM 46 'Making a Clock'**

1 Turn the hands on the clock to show the time 3 o'clock.

a Draw what the angle of the hands looks like.

b How do you know this angle is equal to a quarter turn?

2 Turn the hands to find other times when the hands of the clock show the same angle. Record the times on the clocks.

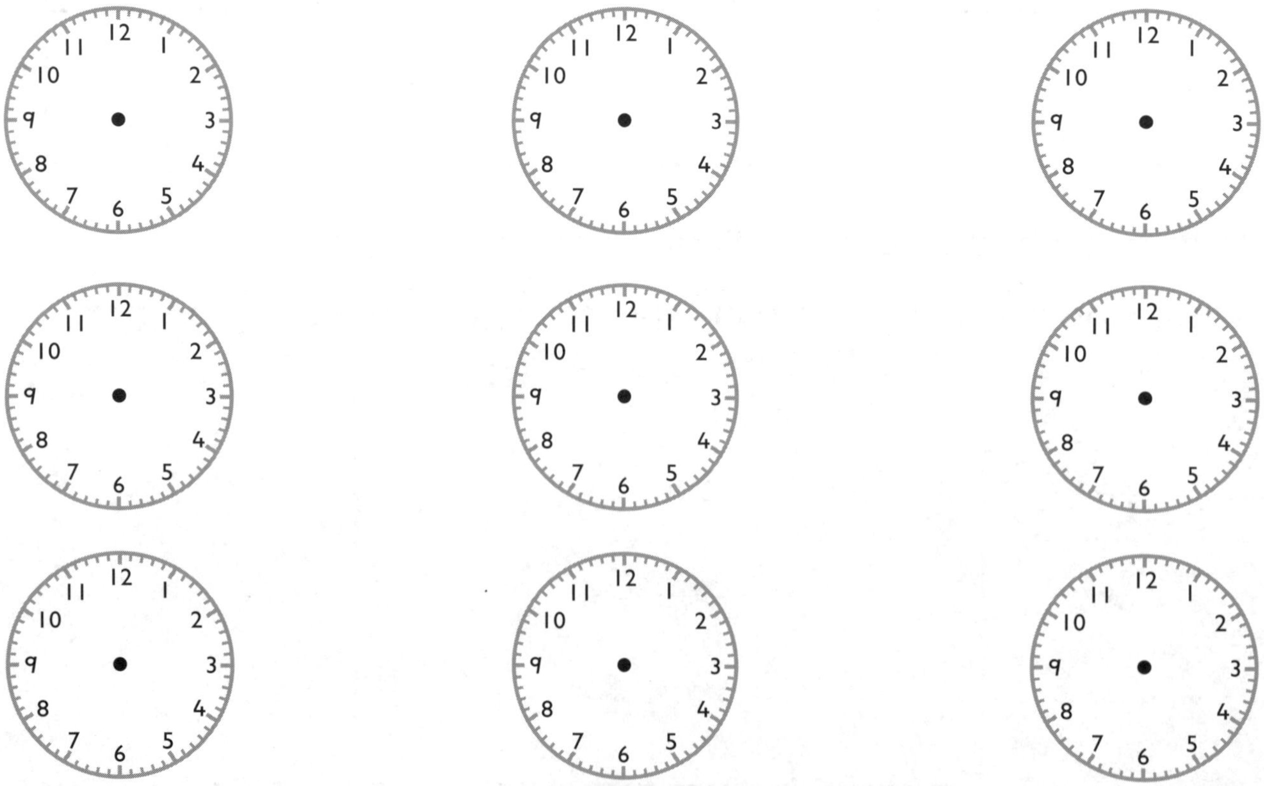

DATE:

STUDENT ASSESSMENT

1 Draw these lines.

curved	vertical	horizontal	diagonal

2 In your own words, describe what an angle is.

____

____

3 Draw and label 4 angles that you can see in the classroom.

4 **a** Your friend is having trouble understanding angles. Explain how you would use a clock to show them an angle that is equal to a quarter turn.

____

____

b Draw a picture of the angle that you would show them.

At the Supermarket

DATE:

1 Complete these skip counting patterns.

a 2, 4, 6, ______, ______, ______, ______, ______, ______, ______, ______

b 5, 10, 15, ______, ______, ______, ______, ______, ______, ______, ______

c 10, 20, 30, ______, ______, ______, ______, ______, ______, ______, ______

2 Luis was at the supermarket.

Luis wanted to know how many of each item he would receive if he bought the following. Record the multiplication problems he should use. Use the skip counting patterns from Question 1 to help you solve them. One has been done.

4 bags of bread rolls 4 × 10 = 40	5 boxes of pizzas	6 bags of apples	3 packets of biscuits
7 cartons of eggs	1 box of soup	2 bags of bread rolls	6 lots of juice
5 boxes of pizzas	8 bags of bread rolls	3 lots of juice	1 bag of apples
4 boxes of soup	5 packets of biscuits	2 cartons of eggs	6 boxes of soup

3 How did skip counting help you solve these multiplication problems?

____

Doubles and Multiplying by 2

DATE:

1 Draw lines matching each times table and doubles fact to its answer. One has been done.

5 × 2	2	double 3
9 × 2	10	double 10
7 × 2	8	double 5
6 × 2	12	double 7
1 × 2	20	double 4
2 × 2	6	double 9
4 × 2	18	double 8
10 × 2	14	double 6
3 × 2	4	double 2
8 × 2	16	double 1

2 Solve these 2-times-table facts. Write the doubles fact you used to help you. One has been done.

a 5 × 2 = 10 ↔ double 5 is 10

b 6 × 2 = ____ ↔ ____________

c 8 × 2 = ____ ↔ ____________

d 4 × 2 = ____ ↔ ____________

e 10 × 2 = ____ ↔ ____________

f 2 × 2 = ____ ↔ ____________

3 Solve these problems as quickly as you can.

×	4	9	2	7	1	3	8	10	5	6
2										

What strategy did you use to help you? ____________

____

Multiply by 1, Multiply by 10

DATE:

1 Write the multiplication fact that matches the MAB blocks. One has been done.

a

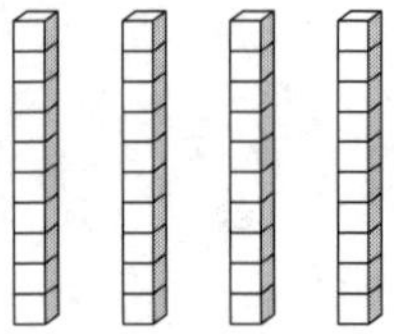

4 × 10 = 40

b

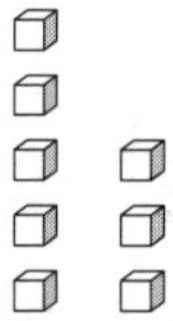

c

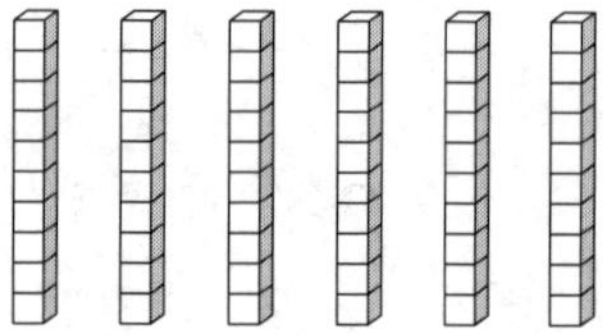

d

e

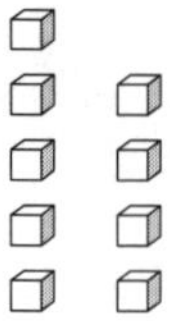

f

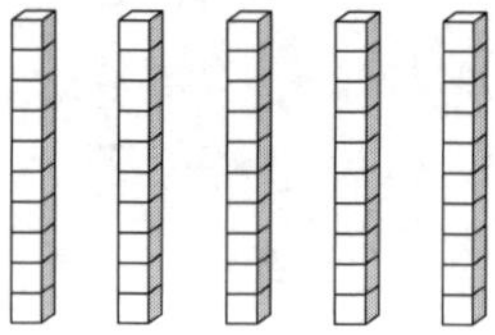

g

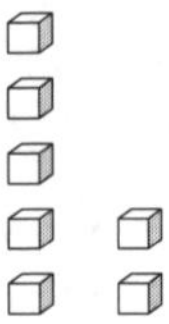

h

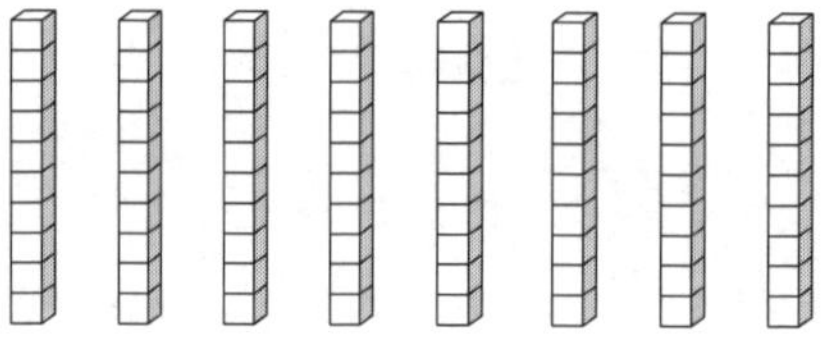

i

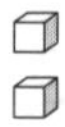

2 Solve these multiplication facts.

a 4 × 1 =	**b** 8 × 1 =	**c** 5 × 1 =	**d** 10 × 1 =
e 7 × 1 =	**f** 2 × 1 =	**g** 6 × 1 =	**h** 3 × 1 =
i 8 × 10 =	**j** 9 × 10 =	**k** 3 × 10 =	**l** 2 × 10 =
m 9 × 10 =	**n** 7 × 10 =	**o** 10 × 10 =	**p** 4 × 10 =

3 What do you notice about the numbers that are multiplied by 1?

4 What do you notice about the numbers that are multiplied by 10?

Unit **18**

DATE:

STUDENT ASSESSMENT

1 Solve the multiplication facts in these wheels.

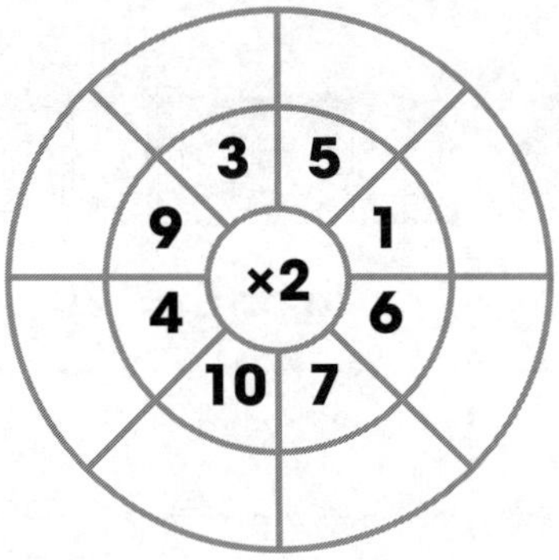

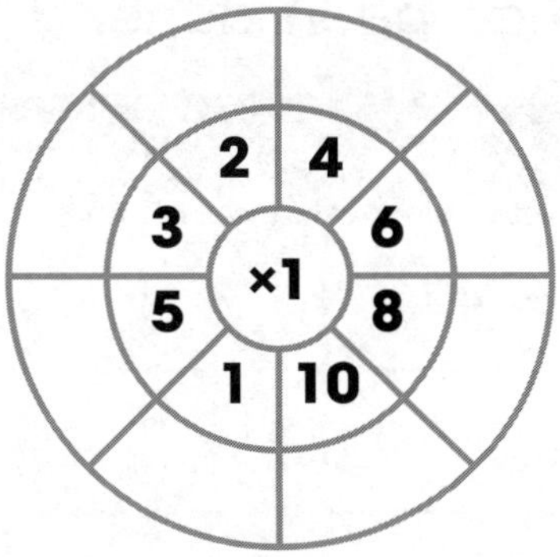

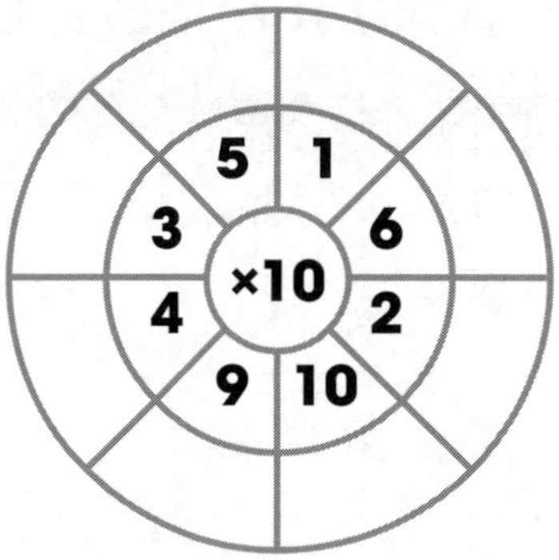

2 Solve these problems. Write the doubles fact you used to solve it. One has been done.

a	5 × 2 = 10	Double 5 is 10.
b	7 × 2 =	
c	9 × 2 =	

3 Draw a line matching the multiplication fact to the MAB blocks.

4 × 10 = ________ 7 × 1 = ________ 9 × 10 = ________

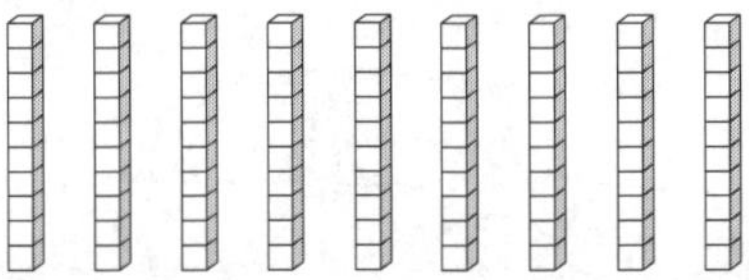

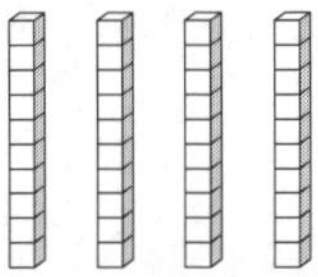

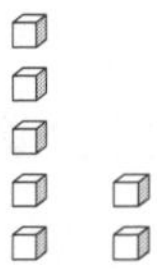

4 Solve each problem. Write the strategy you used.

a 8 × 2 = Strategy: ______________________________

b 4 × 1 = Strategy: ______________________________

c 5 × 10 = Strategy: ______________________________

5 How can skip counting help you solve multiplication problems?

____

Multiply by 5

DATE:

Solve these 5 times table facts by multiplying by 10 and halving.

1 Complete the table. One has been done.

5 times table fact	Multiply by 10	Halve it	So …
4 × 5	4 × 10 = 40	20	4 × 5 = 20
7 × 5			
1 × 5			
3 × 5			
8 × 5			
9 × 5			
2 × 5			
	6 × 10 = 60		
		25	

2 There are 5 seats in a car.

If Silvia saw the following number of cars, how many seats were there?

a 6 cars ____________ **b** 8 cars ____________ **c** 3 cars ____________

d 7 cars ____________ **e** 4 cars ____________ **f** 10 cars ____________

What mental strategy did you use to work out these problems?

____

____

Multiply by 3

DATE:

1 Complete the counting by 3s number sequence.

0, 3, 6, ______, ______, ______, ______, ______, ______, ______, ______

2 Help the frogs get to the rocks on the other side of the pond by jumping on lily pads to make 3-times-table facts. One has been done.

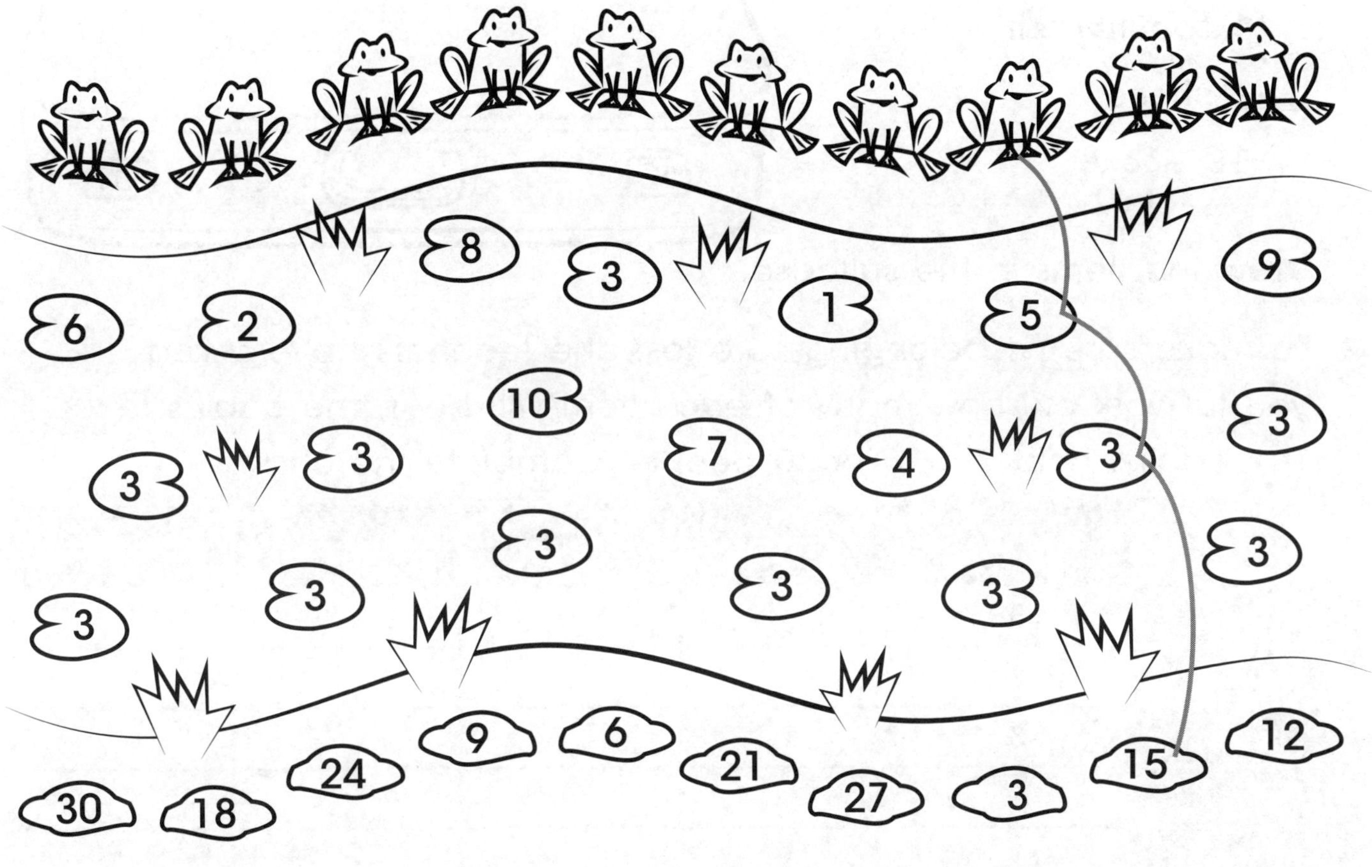

3 What strategy did you use to help get the frogs to the other side of the pond? ______________________________

____

____

Off to Camp

DATE:

1 Imagine you are going on school camp. You need to pack the following items.

8 T-shirts
2 pairs of pyjamas
3 jumpers
7 pairs of shorts
1 toothbrush
4 shoes
10 socks

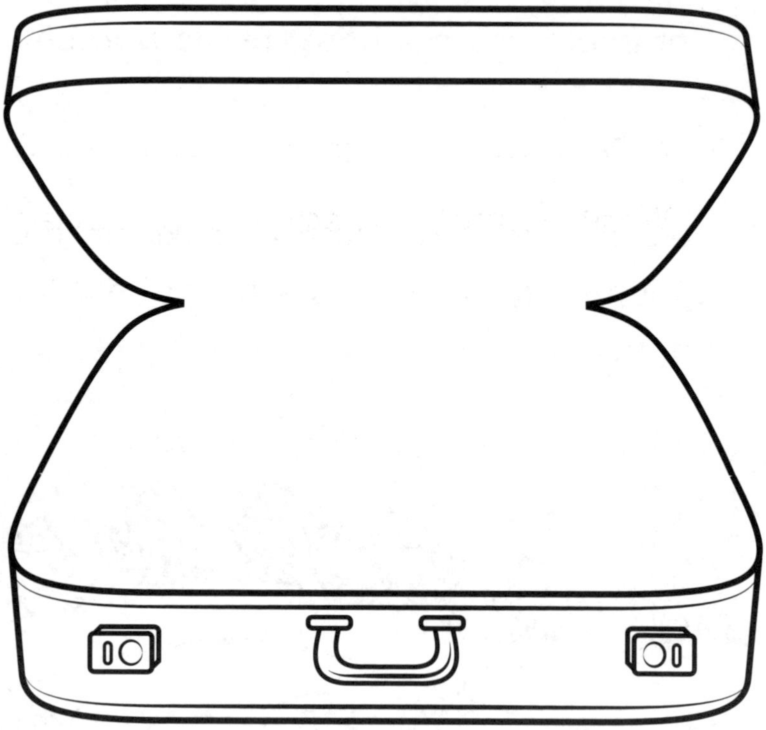

Draw the items in the suitcase.

2 To make sure no belongings are lost, the teachers have asked you to work out how many of each item will be in the cabins. The cabins hold 2, 3, 5 or 10 people. Complete the chart.

	1 person	2 person	3 person	5 person	10 person
T-shirts	8				
pyjamas	2				
jumpers	3				
shorts	7				
toothbrushes	1				
shoes	4				
socks	10				

3 Explain the strategies you used to solve the problem.

Unit 19

DATE:

STUDENT ASSESSMENT

1 Solve these multiplication facts.

a 7 × 5 = **b** 8 × 5 = **c** 3 × 5 = **d** 9 × 5 =

What mental strategy did you use to solve them?

2 Solve these multiplication facts.

a 4 × 3 = **b** 10 × 3 = **c** 2 × 3 = **d** 7 × 3 =

What mental strategy did you use to solve them?

3 Complete the following table.

×	1	2	3	5	10
1					
2					
3					
4					
5					
6					
7					
8					
9					
10					

Write the strategies you used.

What's the Chance?

DATE:

1 Read the statements below. Write a word from the box beside each statement to describe the chance of it happening.

likely	unlikely	even chance	impossible	certain

a You will roll an even number on a dice. ______

b You will have fire-drill practice today. ______

c You will walk home from school today. ______

d You will eat an ice-cream at lunchtime. ______

e You will toss heads on a coin. ______

f Tomorrow will be Friday. ______

g Your teacher will turn into a fish. ______

h The next lesson will be music. ______

2 Write down the statement you think is the **most likely** to happen.

______

Why do you think this?

______

3 Write down the statement you think is the **least likely** to happen.

______

Why do you think this?

______

4 Write down two other things that you think are **unlikely** to happen today.

______

5 Your teacher said that after lunch on Wednesday your class would be **likely** to do something. What might it be?

______

Why do you think that?

______

Spinners

DATE:

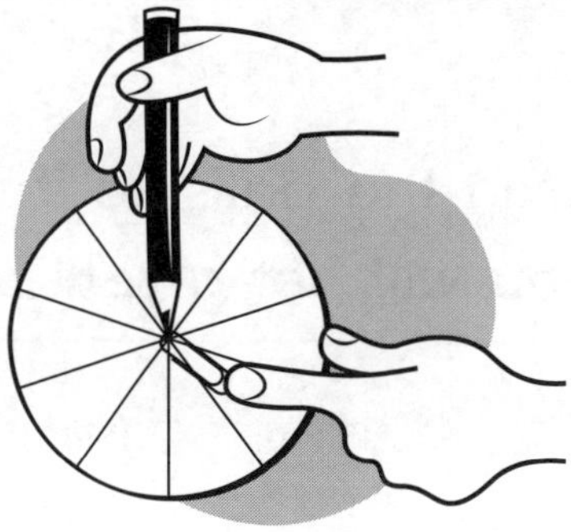

You will need: a partner, a paperclip, a sharp pencil

1 Colour half of the circle green and the other half yellow.

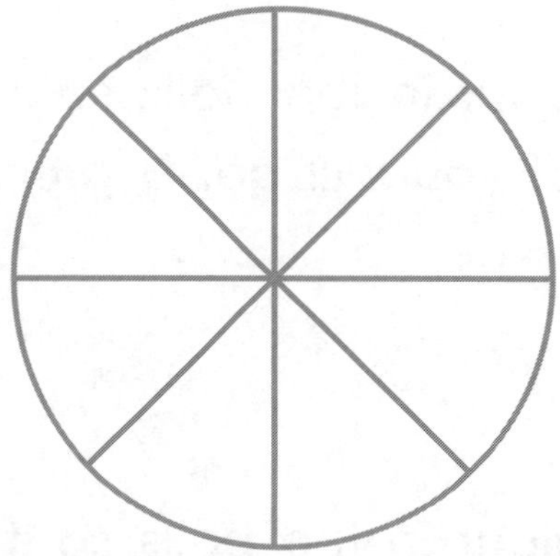

2 If you spin a paperclip 20 times, how many times do you think it will land on yellow? ____________

3 Spin the paperclip 20 times and record the results in the table below.

Green	Yellow

4 Was your prediction correct? ____________

Explain why. ____

____

5 Write your partner's results in the table.

Green	Yellow

6 Were the results the same? ____________

Why do you think that happened?

____

7 What results do you think you might get if you used your spinner 1000 times?

____

Chance Machines

DATE:

1 Colour in the balls so it is **certain** that you will get a **red** ball.

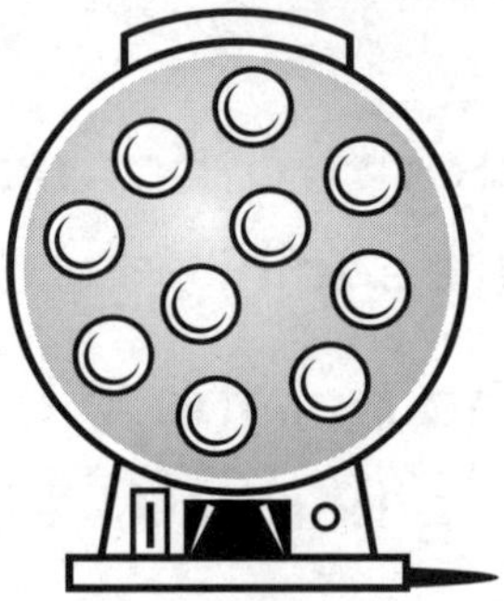

2 Colour in the balls so it is **likely** that you will get a **green** ball.

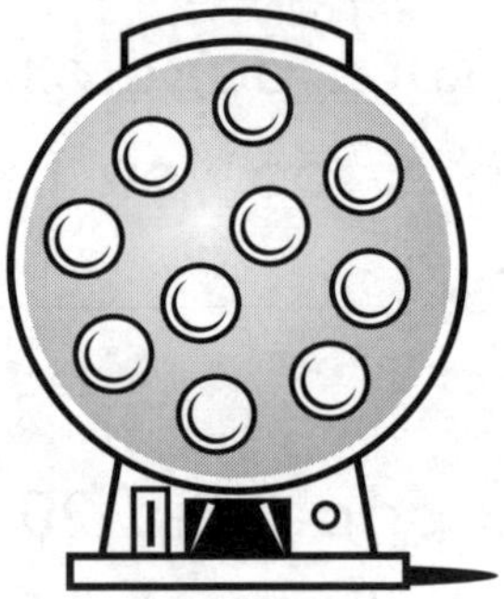

3 Colour in the balls so it is **impossible** that you will get a **yellow** ball.

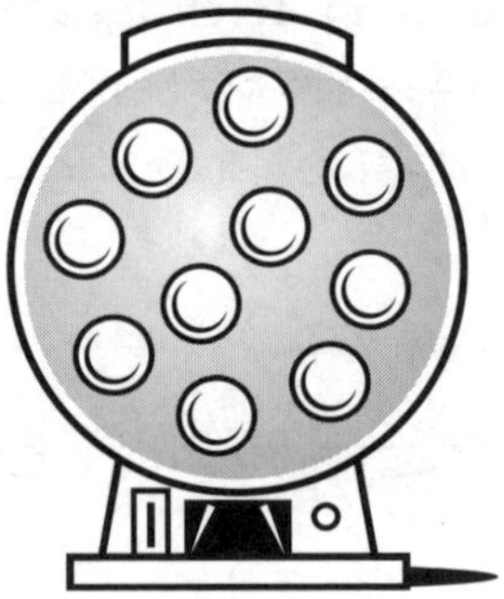

4 Colour in the balls so it is **certain** that you have an even chance of getting a **red** or **blue** ball.

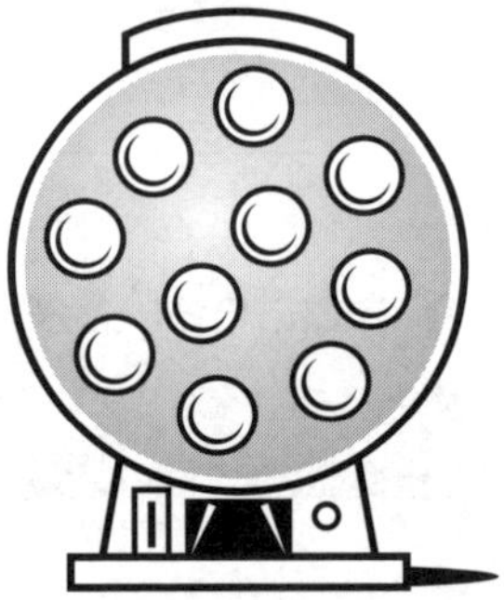

5 Mike puts his money into the toy machine. Do you think he will be happy with the toy he gets? ____________________

Explain why.

____

6 Jo wants a yo-yo. What advice would you give her about spending her money on the machine?

____

Chance (TRB pp. 98–101)
Chance Conduct chance experiments, identify and describe possible outcomes and recognise variation in results (VCMSP147)

DATE:

STUDENT ASSESSMENT

1 Write **likely**, **certain**, **impossible** or **unlikely** to describe the chance of drawing out a black counter from each bag.

______________ ______________ ______________ ______________

2 Colour the spinners to show the following.

There is an **equal** chance of blue and yellow.	It is **impossible** to get black but **likely** to get red.	It is **unlikely** to get red and **likely** to get yellow.	There is a **3 out of 4 chance** of getting green.

3 There were 2 blue, 1 green and 2 yellow counters in a bag. What do you predict will happen after 2 draws from the bag?

____

4 After 20 draws, these were the results.

Blue	Green	Yellow
12	3	5

Are the results the same as you predicted? ______________

Explain why.

____

Jigsaw 100 Chart

DATE:

You will need: **BLM 54 'Jigsaw 100 Chart'**, scissors, glue

Fill in the missing values on the **BLM 54 'Jigsaw 100 Chart'**. Then cut out the pieces and arrange them to form a 100 chart in the grid below.

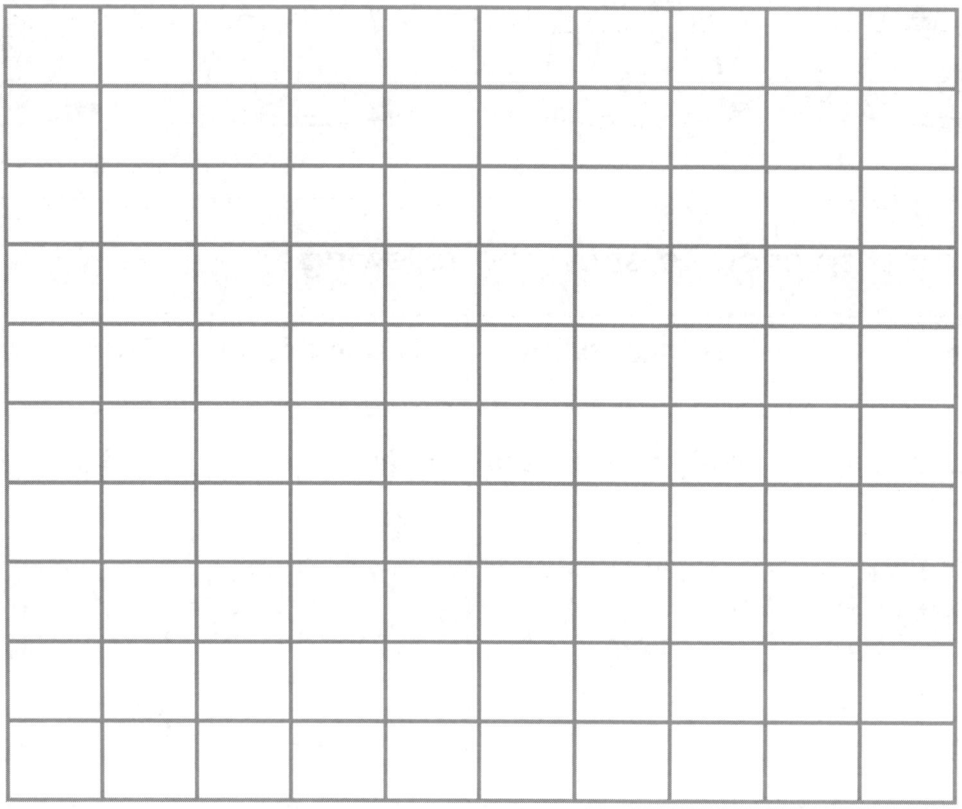

Challenge!

If the value of the ★ on this jigsaw piece is 25, what could the other numbers be?

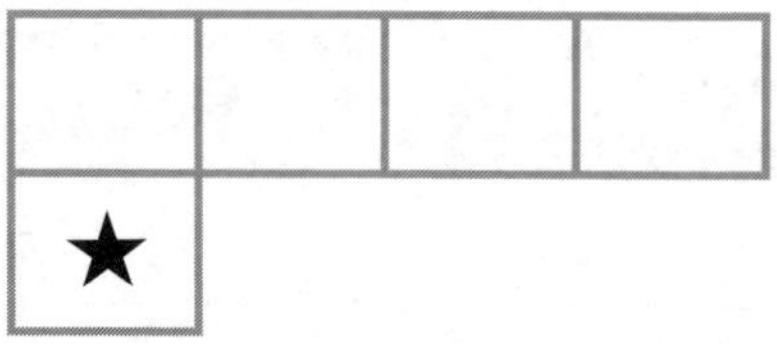

How many possible answers can you think of? Hint: the piece can be rotated to face different ways. Draw your answers on a sheet of paper.

Bounce, Bounce, Stop!

DATE:

Help Kip the Kangaroo complete his Bounce Challenges.

Track them on the number lines.

Challenge 1:

Starting at 3, bounce 7 times. Each bounce will move 4 spaces.

The first one has been done.

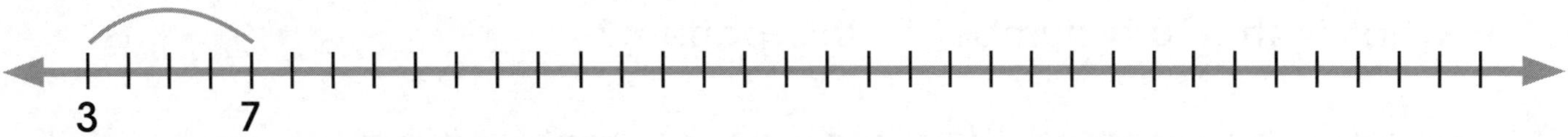

Predict the next 4 numbers in the pattern.

__________ __________ __________ __________

Challenge 2:

Starting at 21, bounce 9 times. Each bounce will move 3 spaces.

Predict the next 4 numbers in the pattern.

__________ __________ __________ __________

Challenge 3:

Starting at 100, bounce 5 times. Each bounce will go backwards 7 spaces.

Predict the next 4 numbers in the pattern.

__________ __________ __________ __________

2-Step Number Patterns

DATE:

1 **a** Complete the number pattern and the jumps on the number line.

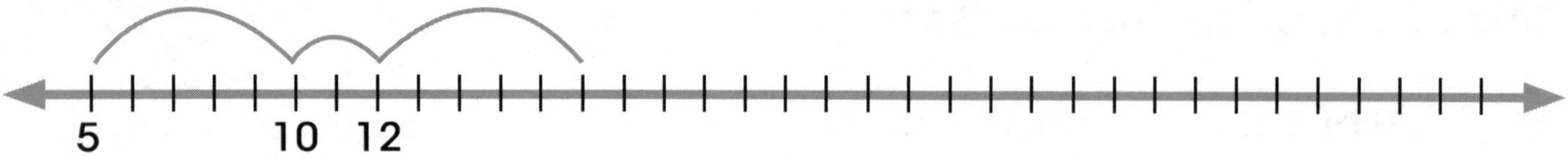

b Write the rule for this pattern. ______________________________

c What is the 10th number in this pattern? ________

2 Continue the patterns. Write the rule for each pattern.

a 14, 13, 17, 16, 20, _____, _____, _____, _____, _____, _____.

Rule: ______________________________

b 90, 83, 85, 78, 80, _____, _____, _____, _____, _____, _____.

Rule: ______________________________

c 2, 4, 3, 5, 4, _____, _____, _____, _____, _____, _____.

Rule: ______________________________

d 3, 9, 19, 25, 35, _____, _____, _____, _____, _____, _____.

Rule: ______________________________

3 **a** Create your own 2-step number pattern.

_____, _____, _____, _____, _____, _____, _____, _____, _____.

b What is the rule?

____

Patterns (TRB pp. 102–105)
Patterns and algebra Describe, continue and create number patterns resulting from performing addition or subtraction (VCMNA138)

DATE:

STUDENT ASSESSMENT

1 **a** On the chart below, starting at 22, colour every second number green.

Write the rule for this pattern. ____________________

b Starting at 22 again, colour every fourth number yellow.

Write the rule for this pattern. ____________________

c How are these patterns similar or different?

____

21	22	23	24	25	26	27	28	29	30
31	32	33	34	35	36	37	38	39	40
41	42	43	44	45	46	47	48	49	50

2 **a** Show the following rule on the number line: Start at 18, add 5.

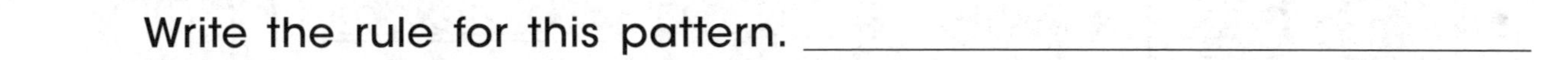

b Predict the next 4 numbers in the pattern.

________ ________ ________ ________

3 Continue the patterns. Write the rule for each pattern.

a 14, 19, 24, 29, 34, 39, ____, ____, ____, ____, ____, ____, ____, ____.

Rule: ____________________

b 24, 25, 33, 34, 42, 43, ____, ____, ____, ____, ____, ____, ____, ____.

Rule: ____________________

c 65, 67, 61, 63, 57, 59 ____, ____, ____, ____, ____, ____, ____, ____.

Rule: ____________________

Sports Room Arrays

DATE:

Find the arrays on the shelves and hooks! Write a multiplication problem and story to match. One has been done.

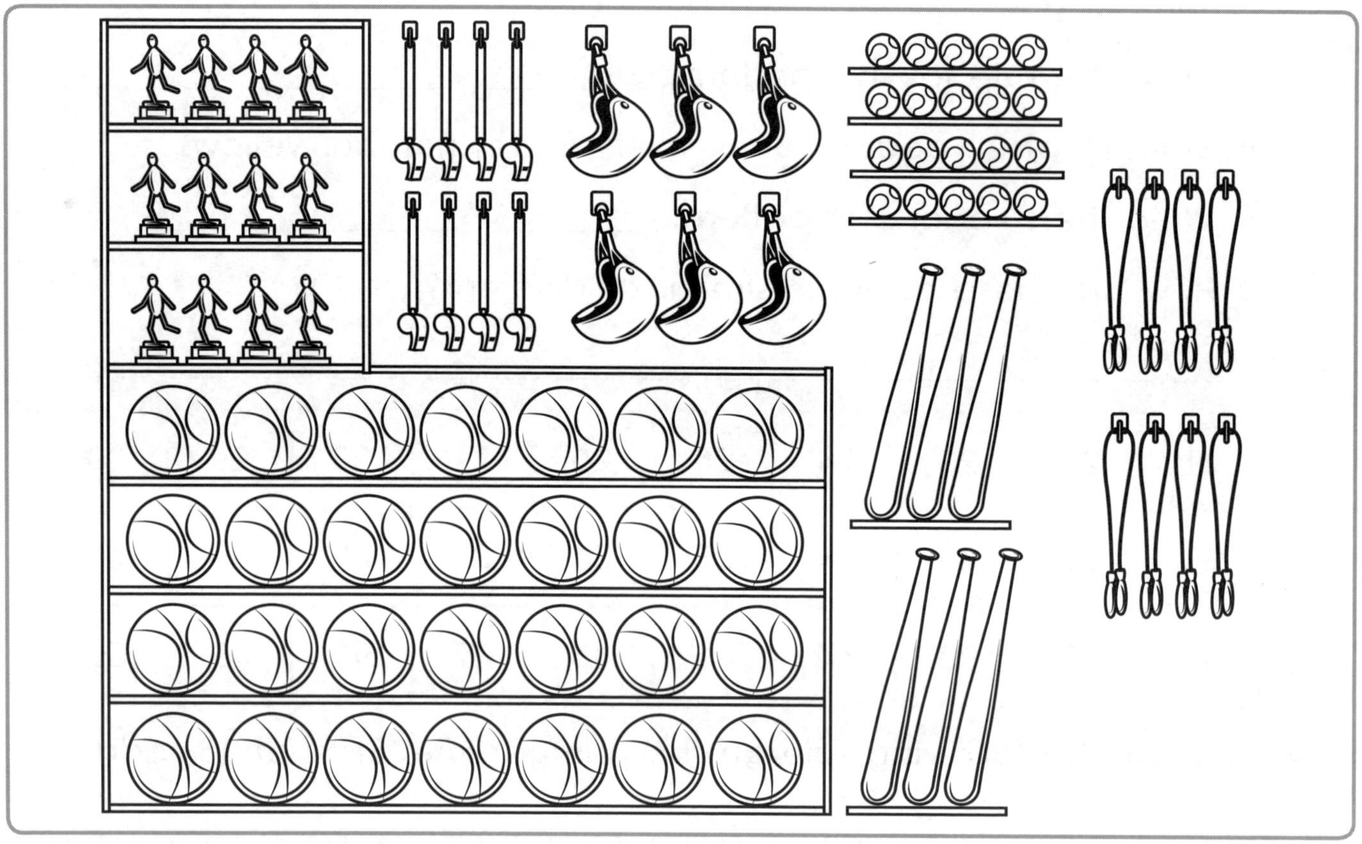

Array	Multiplication problem	Story
	2 × 3 = 6	2 rows of 3 helmets

Multiplication Puzzle

DATE:

You will need: MAB blocks or **BLM 7 'MAB'**

Use MAB blocks to model and solve the **across** and **down** multiplication problems. Write the answers in the grid. One has been done.

Across

1 6 × 6 = 36
3 18 × 4 = ____
4 13 × 5 = ____
6 7 × 10 = ____
7 31 × 2 = ____
9 8 × 2 = ____
10 11 × 11 = ____
11 7 × 5 = ____
12 29 × 2 = ____
14 19 × 4 = ____
15 8 × 8 = ____
17 17 × 5 = ____
18 7 × 9 = ____

Down

2 11 × 6 = ____
3 35 × 2 = ____
5 14 × 4 = ____
6 19 × 4 = ____
8 43 × 5 = ____
9 23 × 5 = ____
11 18 × 2 = ____
13 43 × 2 = ____
14 15 × 5 = ____
16 23 × 2 = ____

Multiplication Problems

DATE:

Solve these worded problems. Set them out vertically.
One has been done.

There were 7 trees in the field with 12 apples on each tree. How many apples were there?	$\begin{array}{r} 12 \\ \times \quad 7 \\ \hline 84 \end{array}$
Jack bought 4 packets of balloons. There were 15 balloons in each packet. How many balloons did he buy?	
Sara planted 5 rows of strawberry plants in her garden. There were 10 plants in each row. How many strawberry plants were there?	
Rufo read 15 pages of his book each day. How many pages did he read in one week?	
There were 12 spiders crawling across the path. How many legs were there?	
Akira baked 12 cupcakes and he put 5 strawberries on top of each one. How many strawberries did he use?	
There were 6 stacks in the corner of the room with 14 boxes in each stack. How many boxes were there altogether?	
Emma visited her grandma on Tuesday, Wednesday and Thursday. Each time she visited her she picked her 23 flowers. How many flowers did she take to her grandma?	

DATE:

STUDENT ASSESSMENT

1 Complete the table. One row has been done.

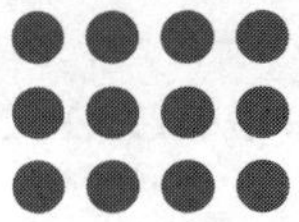	3 rows of 4	3 × 4 = 12
	_____ rows of _____	_____ × _____ = _____
	6 rows of 2	_____ × _____ = _____
	_____ rows of _____	5 × 7 = _____

2 Solve the multiplication problems.

a 16 × 4 = _____

b 23 × 2 = _____

c 15 × 6 = _____

d 26 × 3 = _____

3 Explain how you solved the multiplication problems. _______________

____

4 Solve this problem. There were 4 buses taking the Year 3 children on the excursion. Each bus held 26 children. How many children went?

Area in Playing Cards

DATE:

You will need: playing cards, classroom items

1 Look carefully at the playing cards.

a Use them to **estimate** the area of each object in the table below.

b Then **measure** the area of each object in playing cards.

Object	Estimate	Actual measurement
tabletop		
seat of a chair		
the biggest book		
doormat		
your pencil case		

2 Explain how you estimated the area of each object.

____

____

3 Compare your answers to another student's answers.

Were both answers the same? ________ Why? Why not?

____

____

4 If you wanted to buy a new doormat with **exactly** the same area as the current one, would measuring in playing cards be a good way to find it? ________ Why? Why not?

____

____

Area of Leaves

DATE:

Maria has been collecting leaves from the trees in her garden. She wants to know which leaf has the **largest area**, but is not sure how to work it out. She has traced the leaves onto a grid.

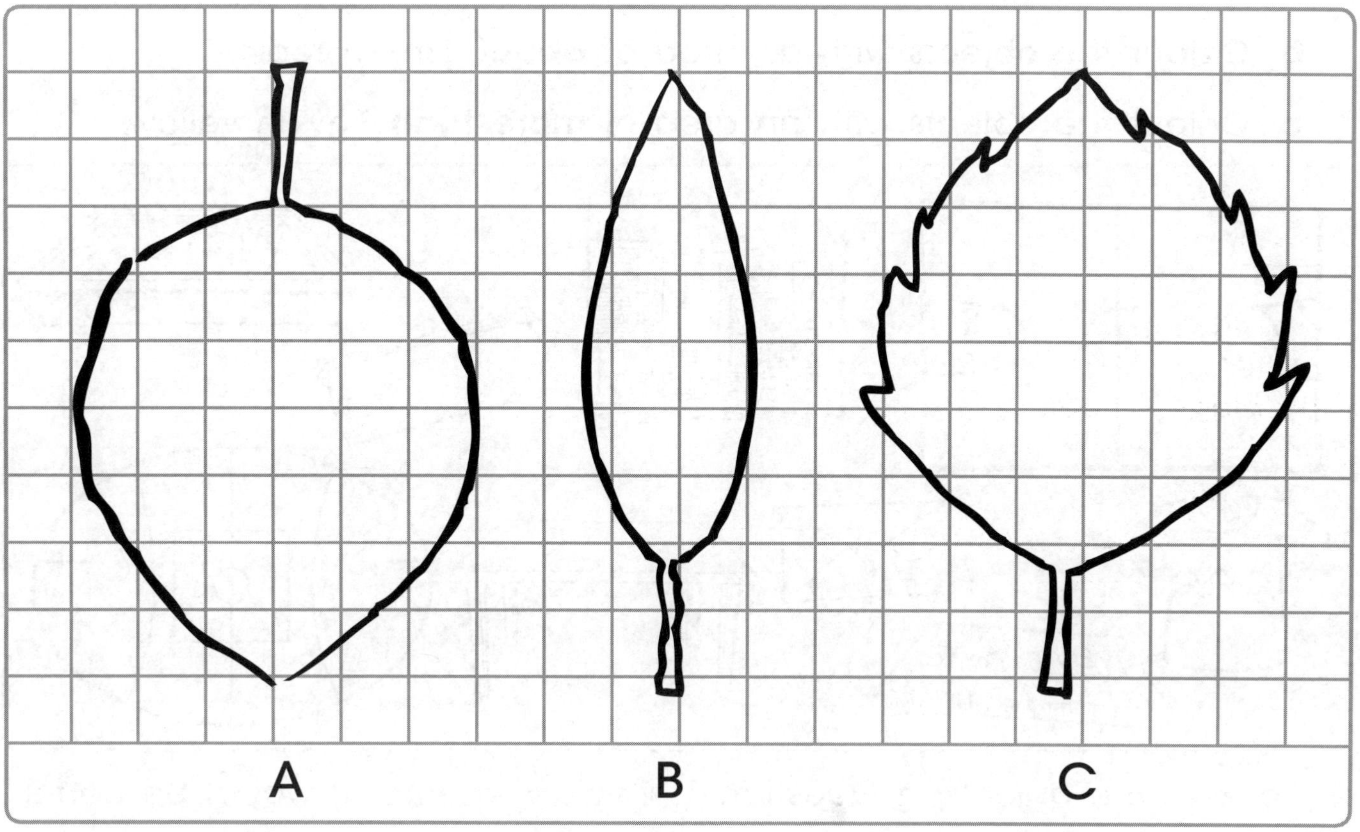

1 Predict the order of the leaves, from **smallest** to **largest**.

____

2 **a** How will you find the area of the leaves? ______________

b What will you do with the whole squares? ______________

c What will you do with the squares that have "more than half" and "less than half" covered? ______________

3 Use your method to find the area of each leaf. Write it inside the leaf as Area = ________ cm^2.

4 Was the order you predicted in Question 1 correct? Yes / No

Square Metre – m^2

DATE:

You will need: coloured pencils

1 Look at the following objects.

a Colour the objects with an area of **less than** 1 m^2 in blue.

b Colour the objects with an area of **about** 1 m^2 in red.

c Colour the objects with an area of **more than** 1 m^2 in yellow.

2 Look at the objects in Question 1. Decide whether it would be better to use square centimetres (cm^2) or square metres (m^2) to measure their area. Write them in the appropriate box.

cm^2	m^2

Unit 23

DATE:

STUDENT ASSESSMENT

You will need: **BLM 61 'Area of Rectangles'**

1 Name the two units of measurement we have used in this unit.

2 Why is it important that everyone measures with the **same** units, rather than some using playing cards, for example, to measure area while others use feet?

3 Look at the rectangles on **BLM 61 'Area of Rectangles'**. What is the **area** of each?

A ______ **B** ______ **C** ______

D ______ **E** ______

4 Draw a rectangle with an area of $12\,cm^2$ on the centimetre grid below.

5 List 2 objects in your classroom that have an area **less than** $1\,m^2$.

6 List 2 objects in your classroom that have an area **more than** $1\,m^2$.

Division Problems

DATE:

You will need: MAB blocks or **BLM 7 'MAB'**

1 Write and solve the following division problems.

a Divide these amounts by 4.

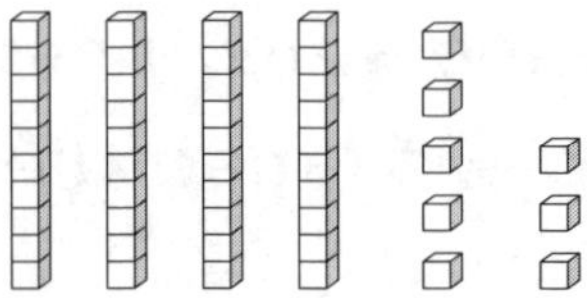

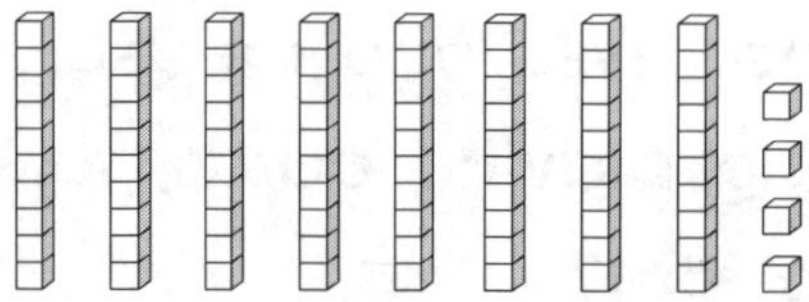

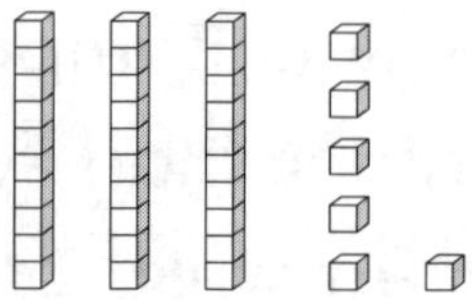

b Divide these amounts by 3.

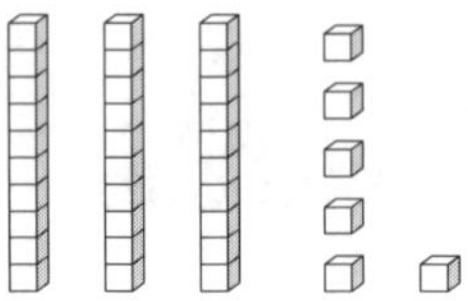

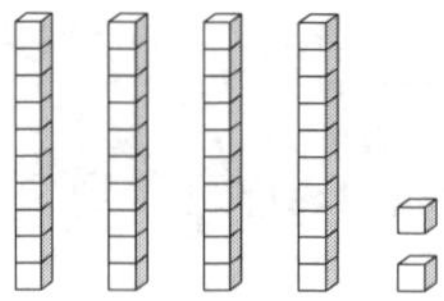

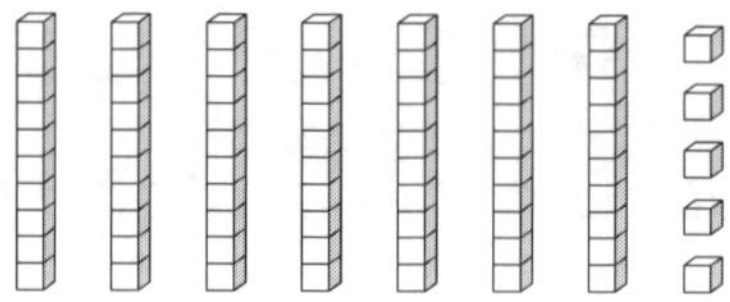

2 Solve these problems using a number from the box. You can use MAB blocks to help you.

a 52 ÷ 4 =   **b** 78 ÷ 3 =   **c** 68 ÷ 2 =

d 85 ÷ 5 =   **e** 84 ÷ 2 =   **f** 93 ÷ 3 =

g 84 ÷ 4 =   **h** 70 ÷ 5 =   **i** 40 ÷ 4 =

42 31 34 14 13 17 26 21 10

Building Blocks

DATE:

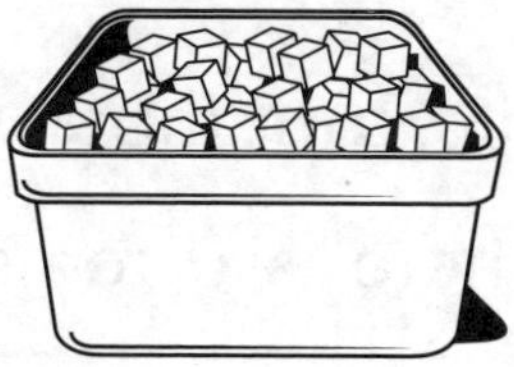

Imagine you have a set of 48 building blocks. You build the objects below, then count how many blocks you used for each one.

1 Work out how many cars you could make using the 48 blocks and record the matching division problem.

2 Repeat for the pyramid, house, aeroplane, rectangular prism, cross and letter "T".

3 Draw 2 other things you could build with the blocks. Record how many blocks they use. Work out how many of each you could build and record the division problem.

car 8 blocks 48 ÷ 8 = ______ 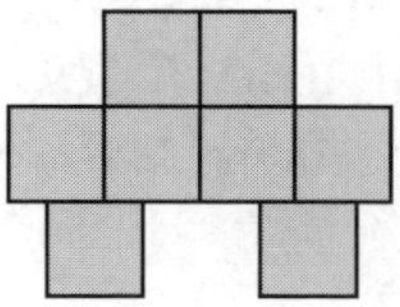	aeroplane 24 blocks 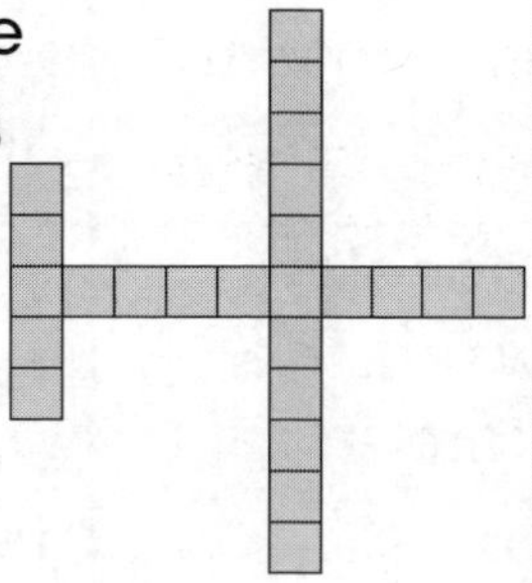	house 12 blocks 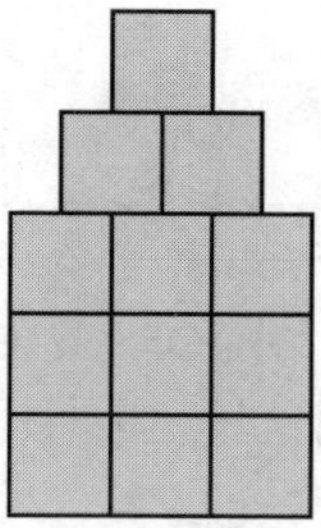
pyramid 6 blocks 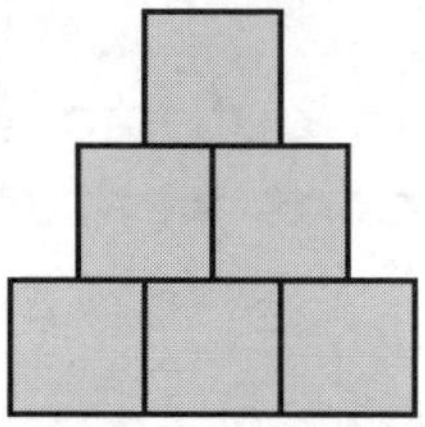	rectangular prism 48 blocks 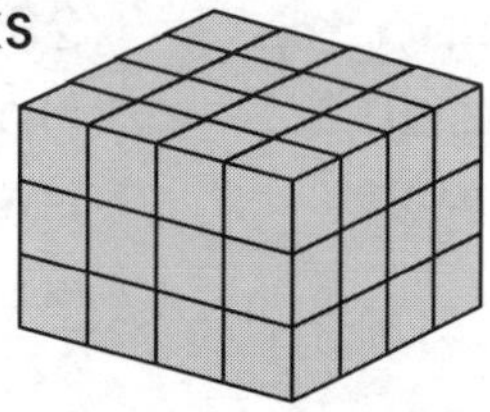	letter "T" 3 blocks 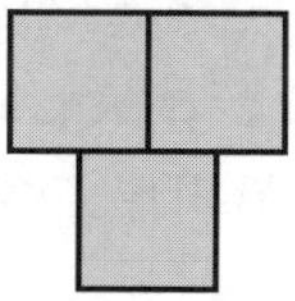
cross 4 blocks 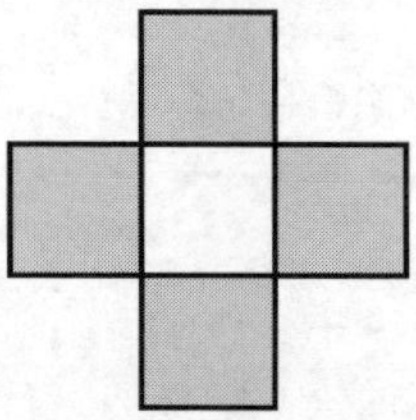	______ blocks	______ blocks

Is There a Remainder?

DATE:

1 Write a number sentence and solve each problem.

If there is a remainder, circle **Yes**. If there is no remainder, circle **No**.

There were 35 children sitting at 4 tables. How many children were at each table? Yes / No	The gardener planted 32 bushes. They were spread out in 6 rows. How many bushes were in each row? Yes / No
25 strawberries was shared between 5 children. How many strawberries did each child get? Yes / No	34 books were placed on 5 shelves. How many books were on each shelf? Yes / No
There are 18 squares of chocolate in a block. There are 3 rows in the block. How many squares are in each row? Yes / No	There were 28 children doing a relay race. They were put into 4 teams. How many children were in each team? Yes / No
There were 48 people travelling by boat. 8 people fit in each boat. How many boats did they use? Yes / No	28 bananas were put into 2 rows. How many bananas were in each row? Yes / No

2 Write a problem that has a remainder.

Then write a problem that does **not** have a remainder.

Yes	No

Unit 24

DATE:

STUDENT ASSESSMENT

1 Solve these division problems by sharing.

a 45 ÷ 5 =

b 24 ÷ 6 =

c 18 ÷ 9 =

d 40 ÷ 4 =

2 Explain the strategy you used to solve these problems.

____

3 Solve these division problems.

a A greengrocer had 24 apples. He packed them into bags of 6. How many bags did he pack? __________

b 32 people went to watch the school play. They were seated in rows of 8. How many rows of people were there? __________

4 Explain the strategy you used to solve these problems.

____

5 Work out the division problems. Circle those with a remainder.

a 35 ÷ 3 =

b 28 ÷ 6 =

c 30 ÷ 10 =

d 24 ÷ 4 =

e 21 ÷ 7 =

f 37 ÷ 5 =

g 22 ÷ 8 =

h 36 ÷ 6 =

Arrays for Division

DATE:

1 Complete the table.

Array	Division fact
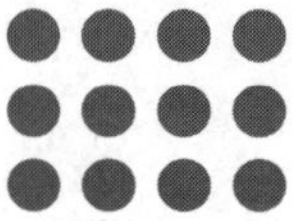	12 ÷ 3 = 4

	30 ÷ 6 =
	24 ÷ 3 =
	28 ÷ 4 =

2 Solve these division problems using an array.

a 18 ÷ 6 =

b 28 ÷ 4 =

c 20 ÷ 4 =

d 25 ÷ 5 =

Division Riddles

DATE:

1 Solve these division riddles by drawing the groups and recording the division problem. You could use multiplication to help you!

a I am a number between 15 and 20. When you divide me into 6 groups there are 3 in each group.

I am ________.

________ ÷ ________ = ________

b I am a number that is smaller than 20. When you share me between 4 groups there are 4 in each group.

I am ________.

________ ÷ ________ = ________

c I am a number that is larger than 20. When you divide me into 4 groups there are 6 in each group.

I am ________.

________ ÷ ________ = ________

d I am a number between 20 and 30. When you divide me into 3 rows there are 7 in each row.

I am ________.

________ ÷ ________ = ________

e I am a number between 25 and 35. When you divide me into 3 rows there are 9 in each row.

I am ________.

________ ÷ ________ = ________

f I am a number that is larger than 30. When you share me between 8 there are 4 in each group.

I am ________.

________ ÷ ________ = ________

2 Write your own division riddle. Give it to a friend to solve!

________ ÷ ________ = ________

Puzzle Pieces

DATE:

The four multiplication and division facts in a fact family fit together like puzzle pieces.

1 Complete these puzzle pieces.

a

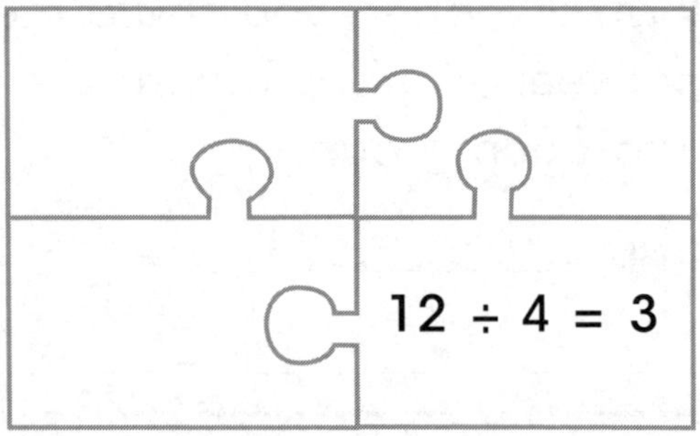

b

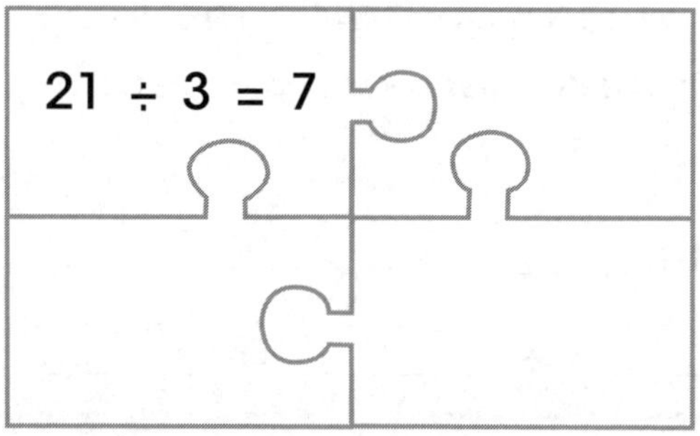

c

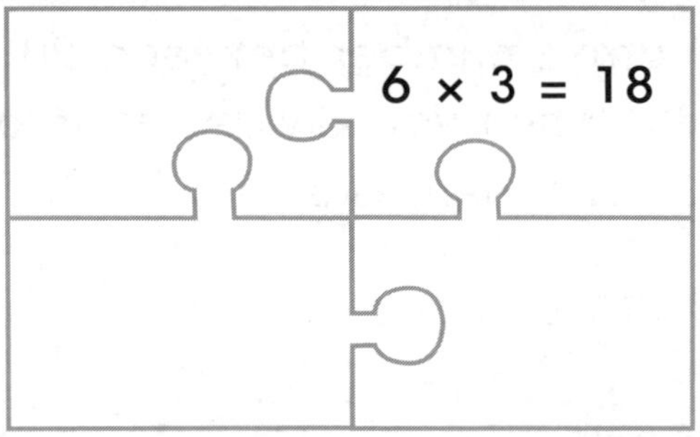

d

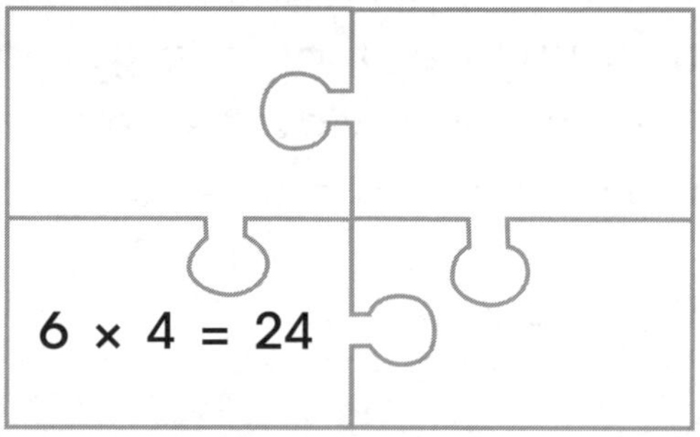

2 Make puzzle pieces using the following numbers.

a 9, 3 and 27

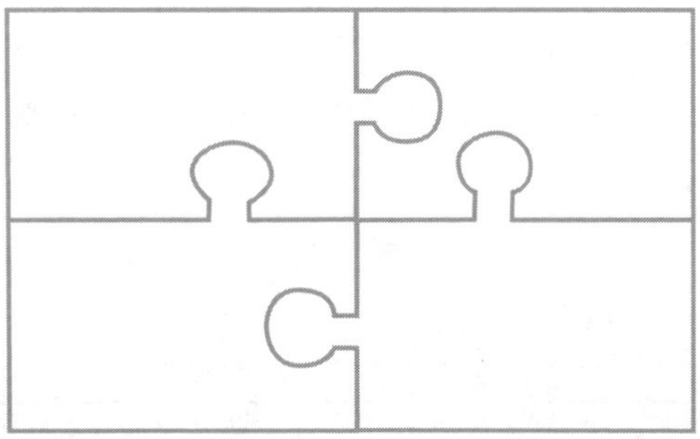

b 4, 5 and 20

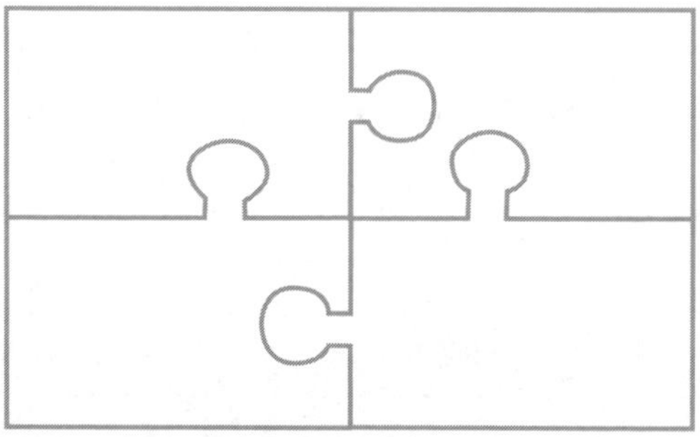

3 Make your own multiplication and division puzzles.

a

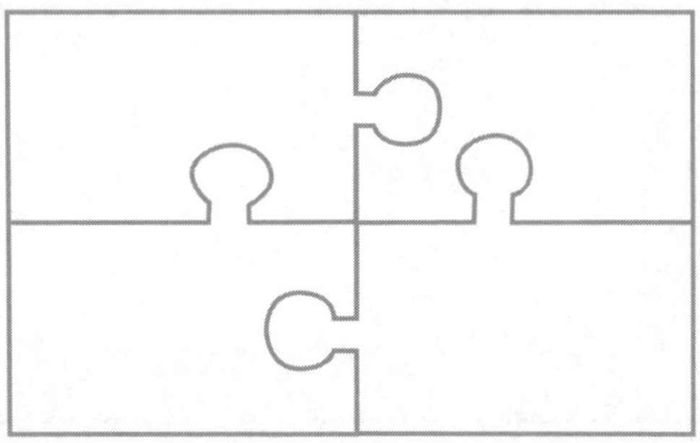

b

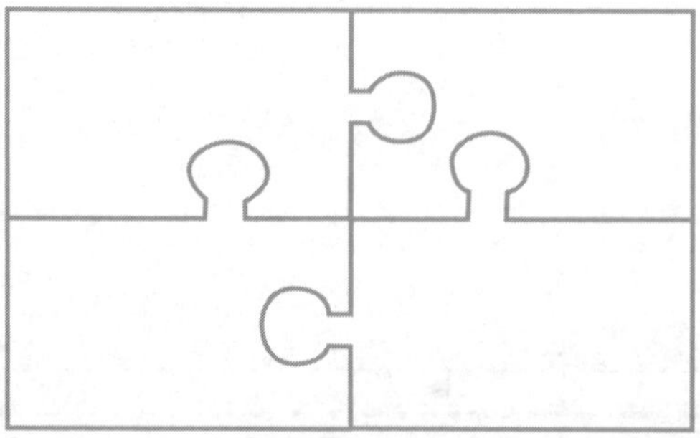

Unit 25

DATE:

STUDENT ASSESSMENT

1 Write a division problem to match these arrays.

a 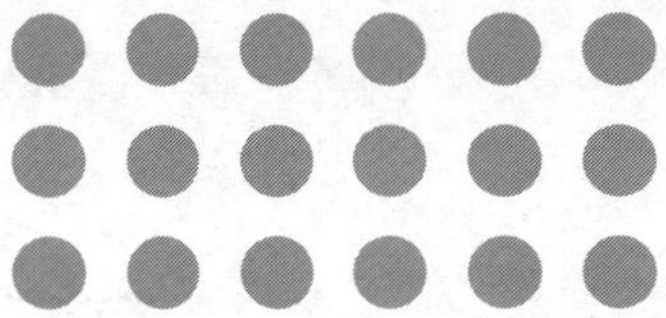

____

b 

____

2 Solve these division problems by drawing an array.

a 25 ÷ 5 =

b 24 ÷ 3 =

3 Explain how you can use multiplication to help you solve this problem: 35 ÷ 5 = ________

____

____

4 Complete these multiplication and division fact families.

a 4 × 5 = 20

20 ÷ 4 = 5

____ × ____ = ____

____ ÷ ____ = ____

b 30 ÷ 10 = 3

10 × 3 = 30

____ × ____ = ____

____ ÷ ____ = ____

c 8 × 2 = 16

____ × ____ = ____

____ ÷ ____ = ____

____ ÷ ____ = ____

5 What four facts could you write using the numbers 3, 5 and 15?

____

____

____

____

Drawing Fractions

DATE:

1 Divide these shapes into 2 equal parts (halves).

2 Divide these shapes into 4 equal parts (quarters).

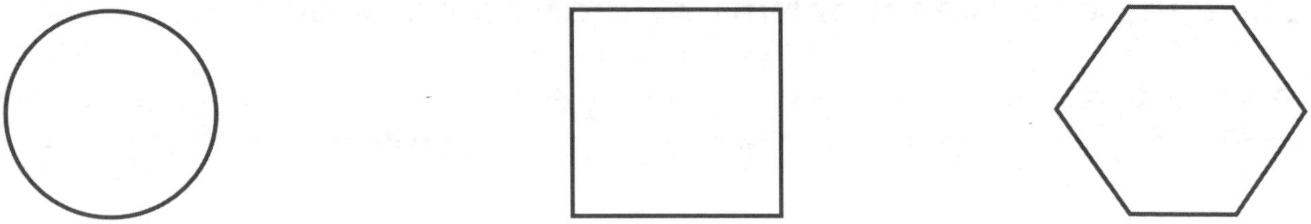

3 Look at the shapes from Questions 1 and 2 to complete this statement:

One half is the same as ______________________________

4 Divide these shapes into 3 equal parts (thirds).

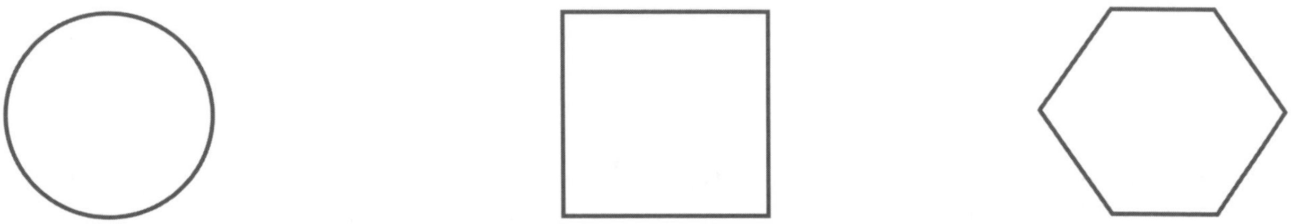

5 Divide each of these shapes into 6 equal parts (sixths).

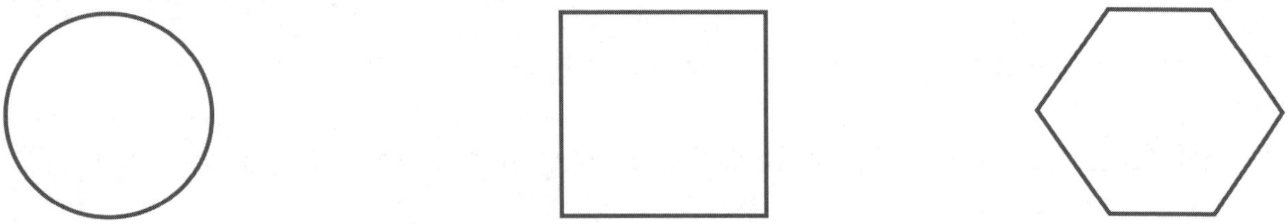

6 Look at the shapes from Questions 4 and 5 to complete this statement:

One third is the same as ______________________________

Fractions 4 Ways

DATE:

Complete this table so that each row has the matching word, fraction symbol and 2 diagrams showing different ways to represent the fraction. One has been done.

Word	Fraction	Diagram 1	Diagram 2
One half	$\frac{1}{2}$		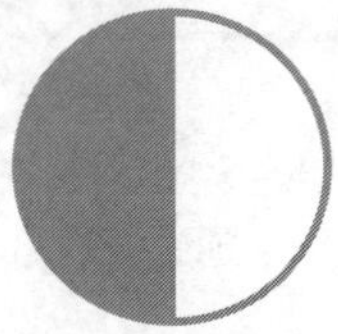
	$\frac{3}{4}$		
Two thirds			
		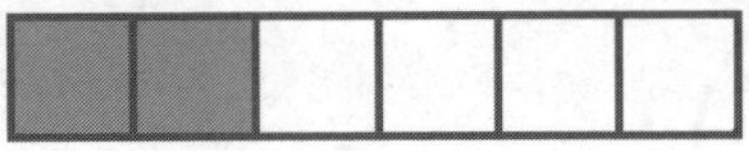	
	$\frac{5}{8}$		
	$\frac{6}{9}$		
Four tenths			

Fraction Shapes

DATE:

1 Write the fraction that represents the shaded part of the whole shape.

a

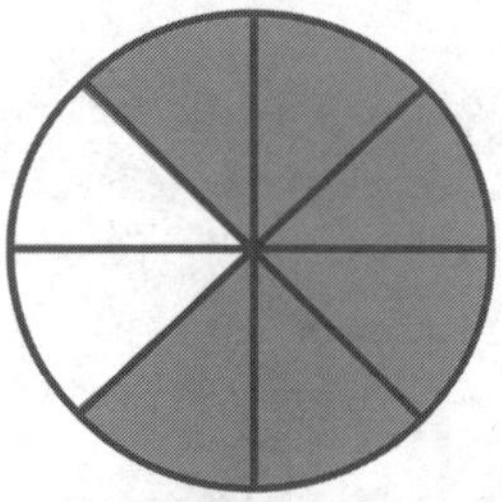

b

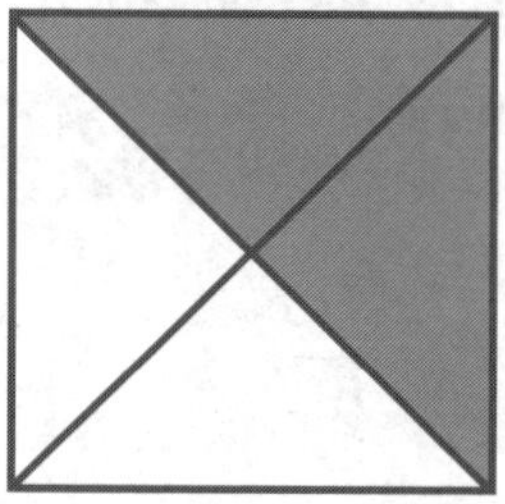

c

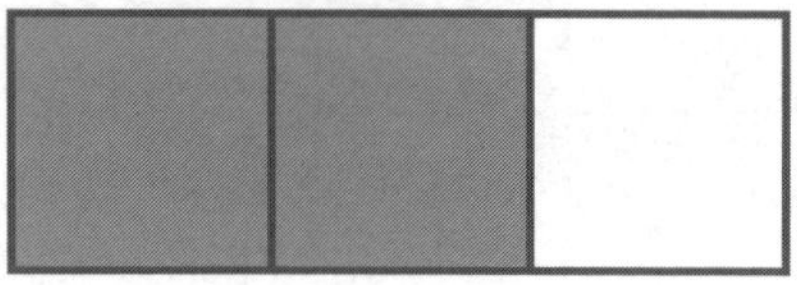

d

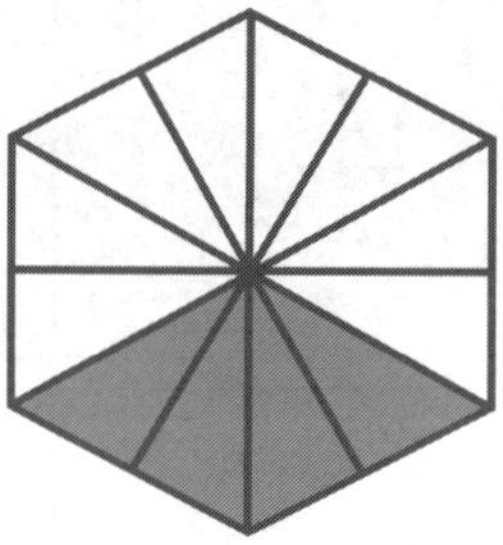

2 Partition the shape and shade parts to match the fraction.

a $\frac{3}{8}$

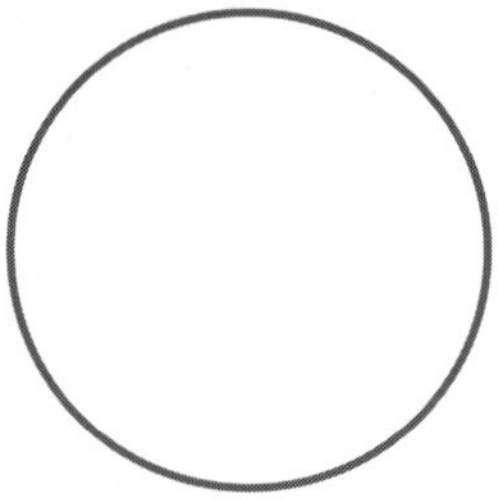

b $\frac{4}{10}$

c $\frac{6}{7}$

d $\frac{2}{6}$

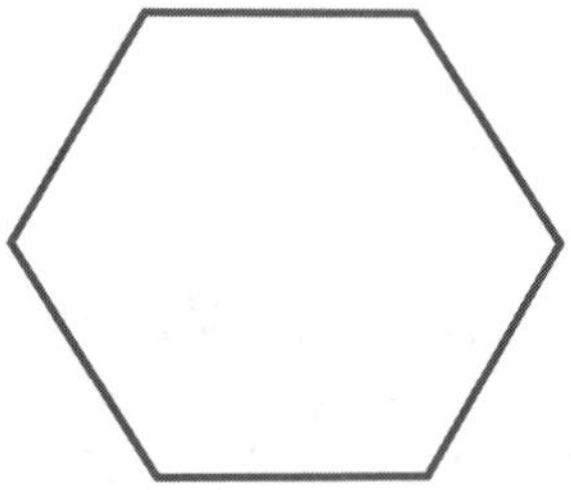

3 For each of the fractions above, circle them:

- **blue** if they are smaller than half.
- **red** if they are equal to half.
- **yellow** if they are bigger than half.

Unit 26

STUDENT ASSESSMENT

DATE:

1 Show 3 different ways to cut these squares in half.

2 Show 2 different ways to cut these squares into quarters.

3 **a** How much bigger is a half than a quarter? ______________

b What does the bottom number on a fraction tell us? ______________

____

c What does the top number tell us? ______________

____

4 Write the fraction that represents the shape shown.

a

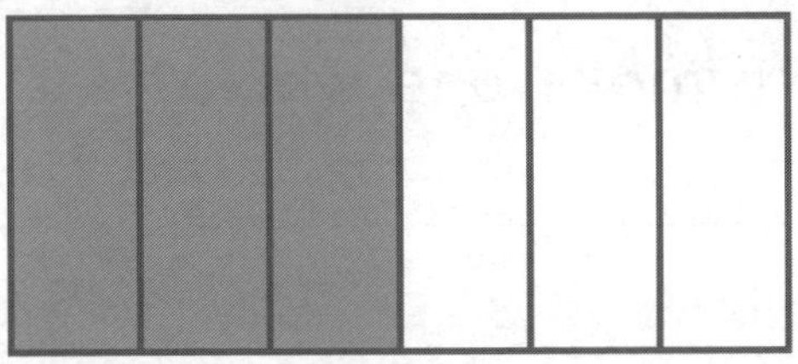

b

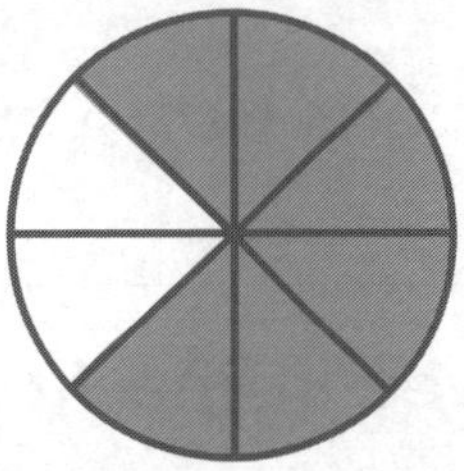

5 Partition the shape. Shade parts to match the fraction shown.

a $\frac{6}{8}$

b $\frac{2}{3}$

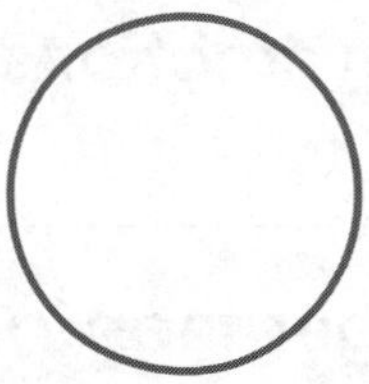

6 Circle the fractions that are **close** or **equal to** a half.

$\frac{4}{5}$ $\frac{3}{8}$ $\frac{2}{9}$ $\frac{58}{100}$ $\frac{10}{12}$ $\frac{3}{10}$ $\frac{4}{7}$ $\frac{2}{4}$

Fraction Wall

DATE:

1 Complete the fraction wall. Use it to answer the questions below.

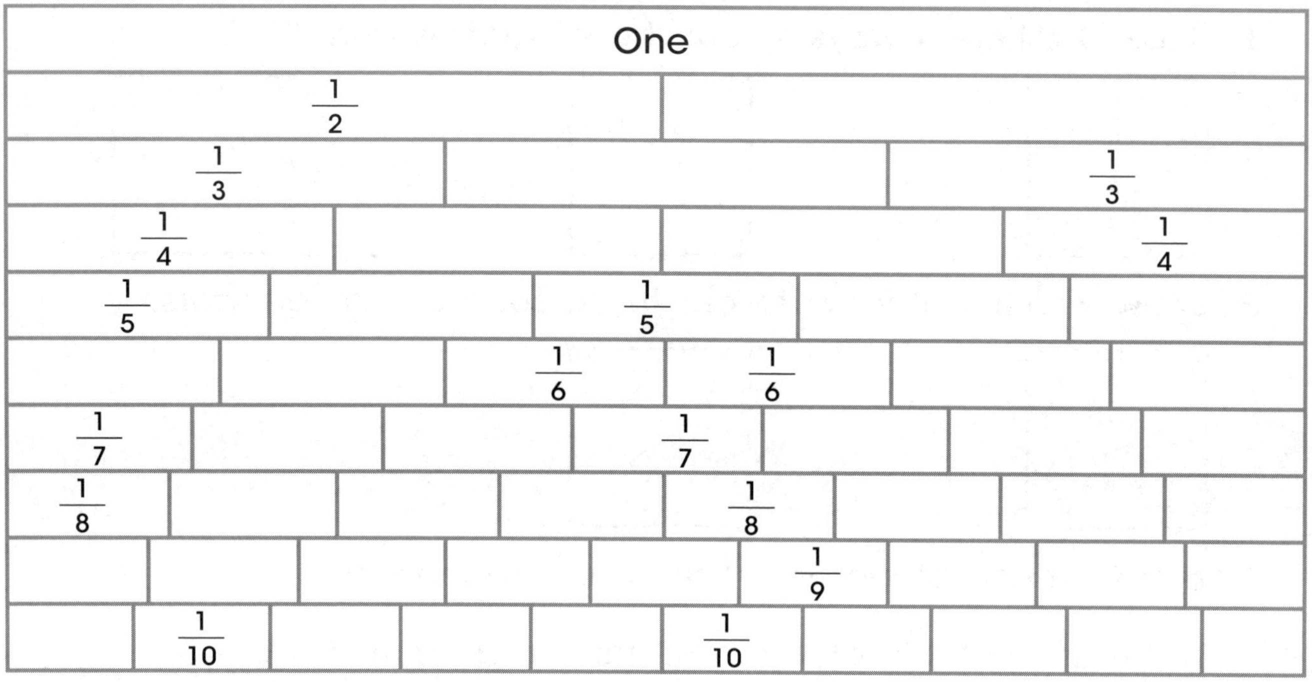

2 **a** How many halves make one whole? ____________

Write this as a fraction: 1 = $\frac{\quad}{2}$

b How many quarters make one whole? ____________

Write this as a fraction: 1 = $\frac{\quad}{4}$

3 What other fractions can you find that make one whole?

____

4 Look at the halves in the fraction wall.

a How many quarters make a half? ____________

b What other fractions make a half?

____

5 Is a half bigger than a quarter? __________

Write other statements about the fractions on the fraction wall.

____

____

Counting by Fractions

DATE:

1 Look for the pattern on each number line and fill in the missing numbers.

a $\frac{1}{2}$ 1 $1\frac{1}{2}$ ☐ ☐ ☐ ☐ ☐ ☐ ☐ ☐ ☐

b $\frac{1}{4}$ $\frac{1}{2}$ $\frac{3}{4}$ ☐ ☐ ☐ ☐ ☐ ☐ ☐ ☐ ☐

c $5\frac{3}{4}$ 6 $6\frac{1}{4}$ ☐ ☐ ☐ ☐ ☐ ☐ ☐ ☐ ☐

d $\frac{1}{5}$ $\frac{2}{5}$ $\frac{3}{5}$ ☐ ☐ ☐ ☐ ☐ ☐ ☐ ☐ ☐

e $\frac{1}{3}$ $\frac{2}{3}$ 1 ☐ ☐ ☐ ☐ ☐ ☐ ☐ ☐ ☐

f ☐ ☐ ☐ $4\frac{1}{4}$ $4\frac{1}{2}$ $4\frac{3}{4}$ ☐ ☐ ☐ ☐ ☐ ☐

2 Circle the **largest** fraction in each group. You may use a fraction wall to help you.

a $\frac{3}{5}$ $\frac{1}{5}$ $\frac{4}{5}$

b $\frac{1}{6}$ $\frac{1}{10}$ $\frac{1}{8}$

c $1\frac{1}{2}$ 2 $\frac{1}{2}$

d $2\frac{1}{4}$ $\frac{3}{4}$ 3

e $\frac{1}{2}$ $\frac{1}{4}$ $\frac{1}{3}$

f $\frac{3}{3}$ $\frac{3}{4}$ $\frac{3}{5}$

Fractions of Collections

DATE:

1 Use the diagrams to answer the problems.

a What is **half** of 14 flowers? ___________

b What is a **quarter** of 24 cars? ___________

c What is a **third** of 27 bananas? ___________

d What is a **fifth** of 45 pebbles? ___________

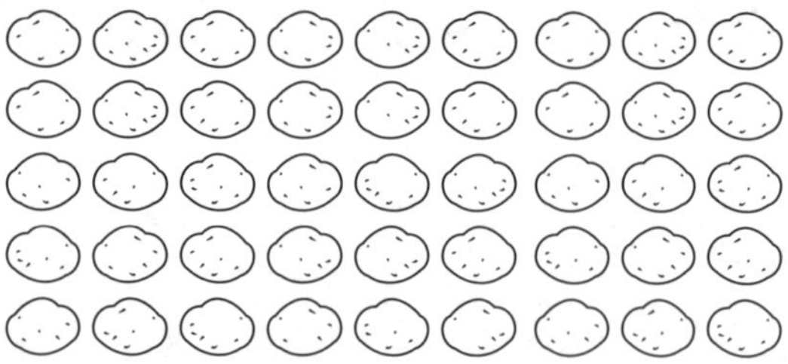

e What is a **tenth** of 20 fish? ___________

2 Explain how you used the diagrams to help you find the answers.

____

3 Using the information above, solve the following:

a What is $\frac{3}{4}$ of 24? ___________

b What is $\frac{2}{3}$ of 27? ___________

c What is $\frac{3}{5}$ of 45? ___________

d What is $\frac{4}{10}$ of 20? ___________

4 How did you use the information from Question 1 to help you?

____

5 Do you know another way to solve these equations? ___________

____

Unit **27** STUDENT ASSESSMENT

DATE:

1 How many thirds make one whole?

Write this as a fraction: $1 = \frac{\quad}{3}$

2 Partition the circle to help answer the following:

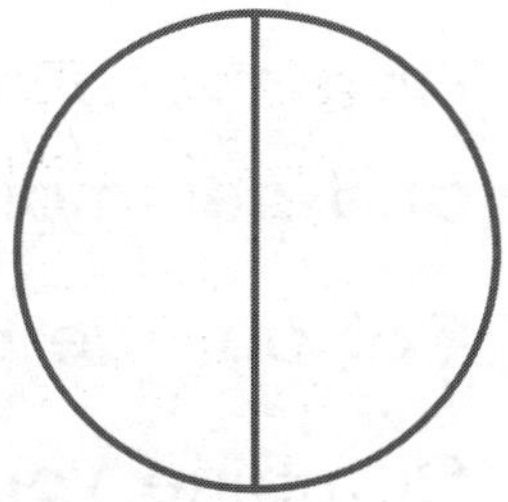

a What fraction is exactly half of one half? ☐

b What fraction is exactly half of one quarter? ☐

3 Look for the pattern on each number line and fill in the missing numbers.

a

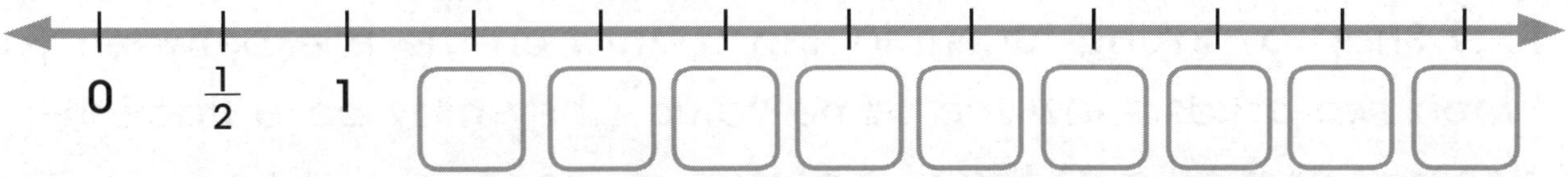

b

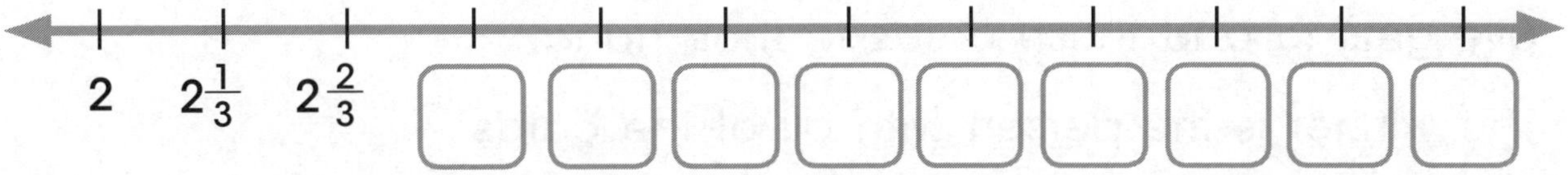

c

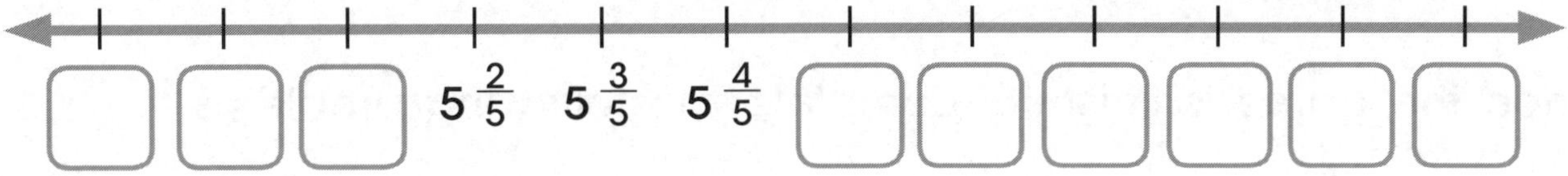

4 Which fraction is bigger: $\frac{1}{3}$ or $\frac{1}{5}$? __________

Explain how you know. ____

____

5 **a** What is one quarter of 16 apples? __________

b Explain how you worked this out.

____

6 **a** What is $\frac{3}{4}$ of 16? __________

b Explain how you worked this out. ________________________________

____

Fraction–Decimal Snap

DATE:

You will need: a partner, **BLM 74 'Fraction–Decimal Cards'**, scissors

How to play

- Fill in the blank spaces on the chart on **BLM 74 'Fraction–Decimal Cards'**. Make sure you shade the diagrams to match the fractions and decimals.
- Cut out the cards.
- Combine your cards with a partner's cards and shuffle.
- Deal the cards so each player has half of the deck.
- Play Snap by taking turns placing a card on the pile between you. When two cards show the same value (they may be a fraction, decimal or diagram) the first person to "snap" their hand over the pile gets to add these cards to their hand.
- The winner is the person with all of the cards at the end of the game.

Once the game is finished, complete the following problems.

1 Convert these fractions to decimals.

a $\frac{5}{10}$ is the same as ______ **b** $\frac{2}{10}$ is the same as ______

c $\frac{7}{10}$ is the same as ______ **d** $\frac{10}{10}$ is the same as ______

2 What decimals are shown by these diagrams?

a 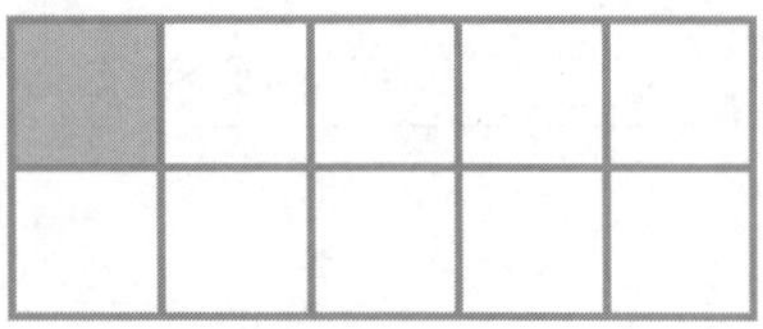 = ________

b 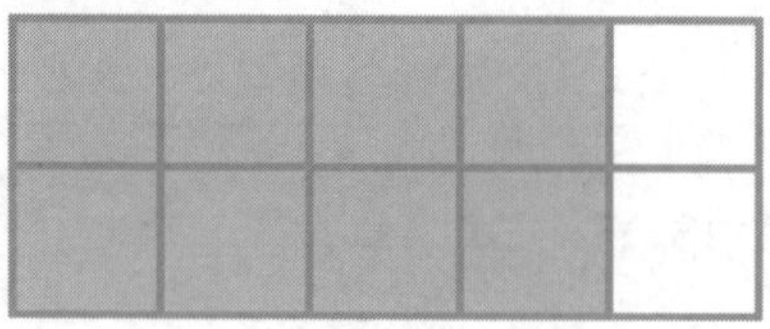 = ________

Zero's the Go! Decimals

DATE:

You will need: 3 friends, playing cards with picture cards, jokers and 10s removed, a calculator

AIM: To be the first to reduce your decimal number to zero without going out!

- Select one person to be the dealer. The other three people are the players.
- The dealer shuffles the deck and reveals four cards, laying them in the order that they are drawn from the deck, to form a 3-digit number with one decimal place. This is the starting number.
- Players write this number on their game board and insert the value into their calculator.
- The dealer draws cards from the deck, one at a time.
- Players decide whether they will use or skip the card. If they decide to use the card, they can choose the place value of the digit and take this away from their number. For example, if a 6 is drawn, it could be worth 600, 60, 6 or 0.6.
- Players record their progress on the game board and calculator.
- The first player to zero wins! (But players are out of the game if they go below zero.)

	H	T	O	•	Tenths
Start ▶					

Number Lines With Decimals

DATE:

You will need: a partner, a stopwatch, **BLM 78 'Human Reaction Time'**

1 Complete each number line pattern and name the counting pattern.

a 0.5 0.6 0.7 ☐ ☐ ☐ ☐ ☐ ☐ ☐ ☐ ☐

The counting pattern is: ______________________

b 3.3 3.4 3.5 ☐ ☐ ☐ ☐ ☐ ☐ ☐ ☐ ☐

The counting pattern is: ______________________

c 2.9 3.1 3.3 ☐ ☐ ☐ ☐ ☐ ☐ ☐ ☐ ☐

The counting pattern is: ______________________

d ☐ ☐ ☐ 1.2 1.4 1.6 ☐ ☐ ☐ ☐ ☐ ☐

The counting pattern is: ______________________

2 Have a partner time you (5 or more times) completing the task on **BLM 78 'Human Reaction Time'**. Record your times in seconds to one decimal place, on **BLM 78**. Then, add markers and labels to the number line to show your times as accurately as possible.

a What was your **slowest** time? ______________

b What was your **fastest** time? ______________

c What is the **difference** between your **slowest** and **fastest** times?

____

DATE:

STUDENT ASSESSMENT

1 Complete the table.

$\frac{5}{10}$		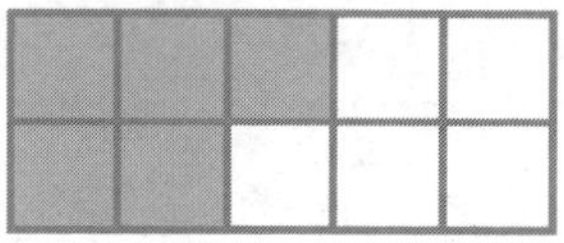
	0.9	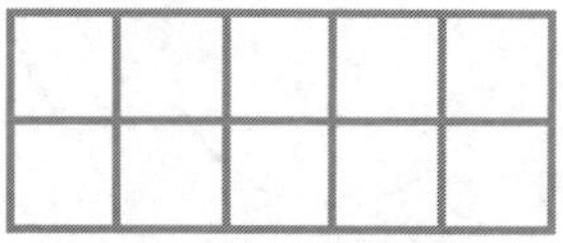
$\frac{4}{10}$		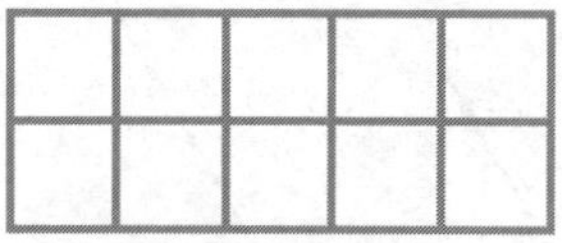
		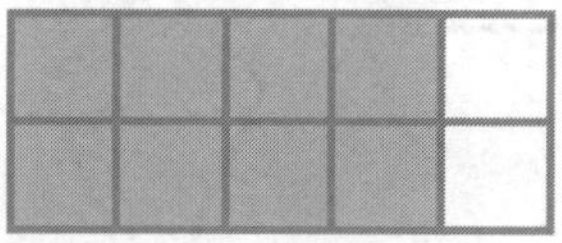

2 Select 4 of the numbers from the box to write on the place-value mat.

9.7 10.4 7.3 8.2 9.1 11.8 8.6 6.9

H	T	O	•	Tenths

3 Write all of the numbers in the box in order from **smallest** to **largest.**

____

4 Place the numbers from Question 2 to scale on the number line.

Describing 2D Shapes

DATE:

1 Colour the squares blue, the triangles green, the circles yellow, the rectangles orange, the hexagons purple, the trapeziums brown, the kites red and the rhombuses black.

2 Draw a shape to match each description.

I have 4 edges and 4 corners. All my edges are the same length.	I have 6 edges and 6 corners. All my edges are the same length.
I have 4 edges and 4 corners. Two of my sides are diagonal lines. My sides are not the same length.	I have no corners and no straight edges.

3 Describe a rectangle. ____

____

Symmetry

DATE:

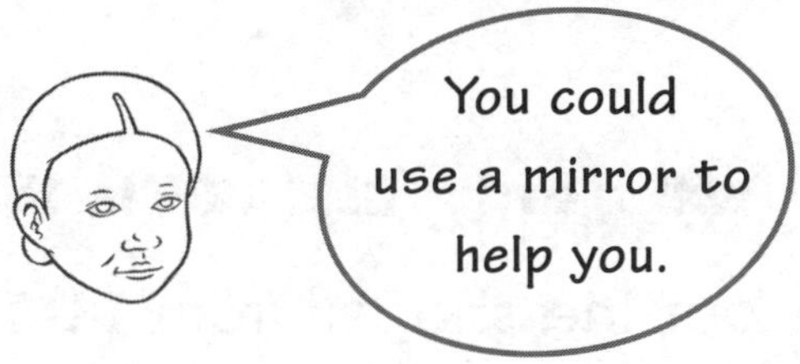

1 Colour the shapes that are symmetrical. Draw the line of symmetry.

2 Complete these symmetrical pictures.

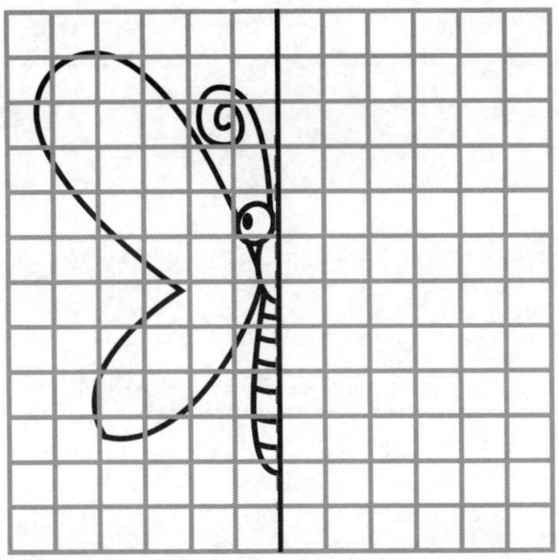

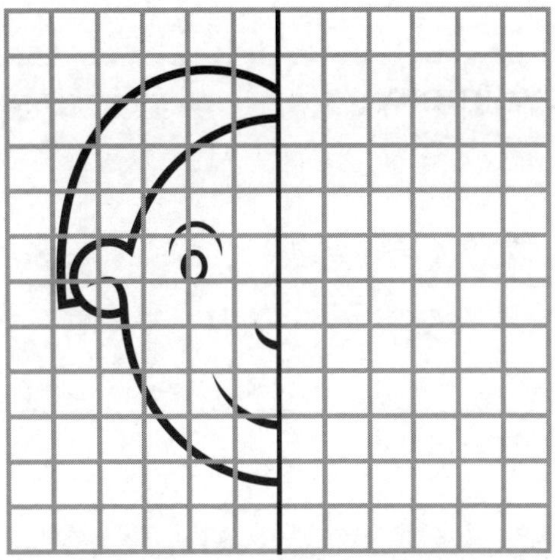

3 Draw one half of a picture on one side of the line of symmetry. Have a partner draw the symmetrical half for you!

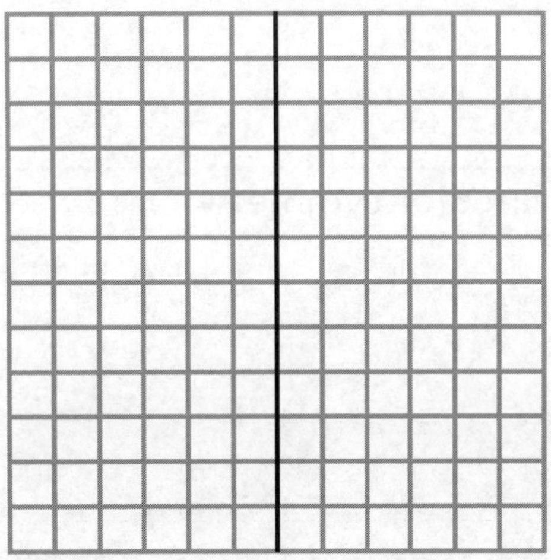

Lines of Symmetry

DATE:

You will need: **BLM 80 'Is It Symmetrical?'** (reduced copy), scissors, glue

Cut out the shapes from **BLM 80 'Is It Symmetrical?'**. Work out how many lines of symmetry the shapes have. Paste them below in the correct box.

no lines of symmetry
1 line of symmetry
2 lines of symmetry
4 lines of symmetry
5 or more lines of symmetry

Unit **29** DATE:

STUDENT ASSESSMENT

1 Write a description for each shape.

a 

____
____
____

b 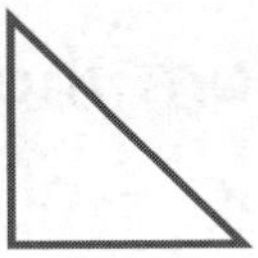

____
____
____

c 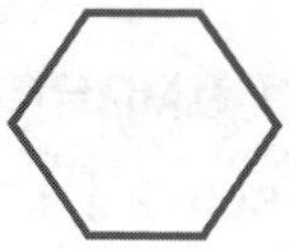

____
____
____

2 Carla has drawn the line of symmetry on some shapes.

a Circle the ones she got correct.

b Now draw the correct line of symmetry on the shapes that Carla got incorrect.

3 Complete these shapes so they are symmetrical.

a 

b 

4 How many lines of symmetry do these shapes have? Draw the lines of symmetry to help you.

a 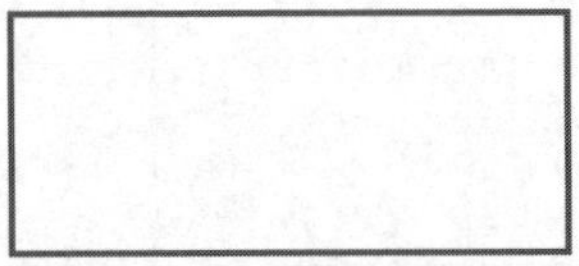

b 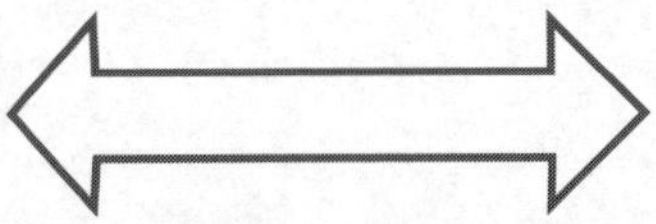

c 

____ lines of symmetry ____ lines of symmetry ____ lines of symmetry

Tin Cans and Egg Cups

DATE:

You will need: an egg cup, a tin can, assorted containers, sand, rice or water

1 Find 4 containers in your classroom that you can find the capacity of. Make sure they are not too big - check with your teacher first.

2 Draw a diagram of each container in the table below.

3 Look at the size of each container and compare it to the tin can. Estimate how many tin cans of sand, rice or water it will take to fill each container. Make sure you write down the material you will use.

4 Now measure how many times you have to refill the tin can to fill each container.

5 Look at the size of the container compared to the egg cup. Repeat steps 3 and 4.

Material used: ____________________

Container		Tin cans		Egg cups	
No.	Diagram	Estimate	Actual	Estimate	Actual
1					
2					
3					
4					

6 What strategies did you use to make your estimates? ____________________

____

Capacity (TRB pp. 138–141)
Using units of measurement Measure, order and compare objects using familiar metric units of length, area, mass and capacity
(VCMMG140)

One Litre

DATE:

You will need: A supermarket catalogue or **BLM 82 'Supermarket Capacity'** (reduced copy), scissors, glue

1 Look through the supermarket catalogue and cut out as many items as you can that measure capacity.

2 Paste each item into the correct box below.

Less than 1 litre	Equal to 1 litre	More than 1 litre

3 Which box contains the most items? ______________ Why do you think this is?

____

4 Can you combine any of your items in the "less than 1 litre" box to make exactly 1 litre? List as many different combinations as you can. Note: you can use multiples of a single item.

____

____

Measuring Jugs

DATE:

You will need: assorted containers, measuring jugs, water

1 How much liquid is in each container?

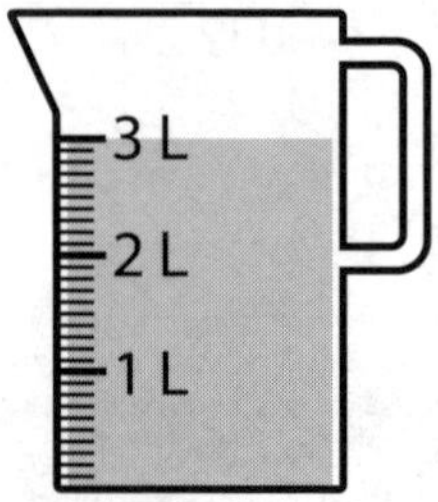

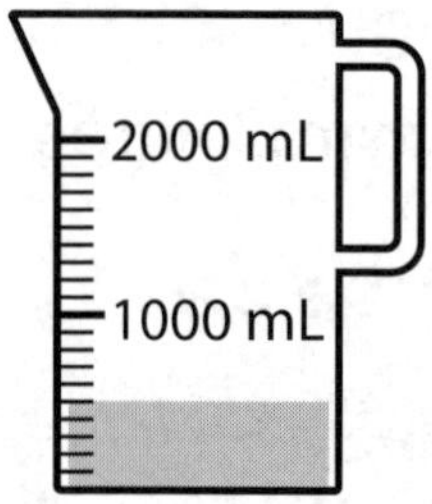

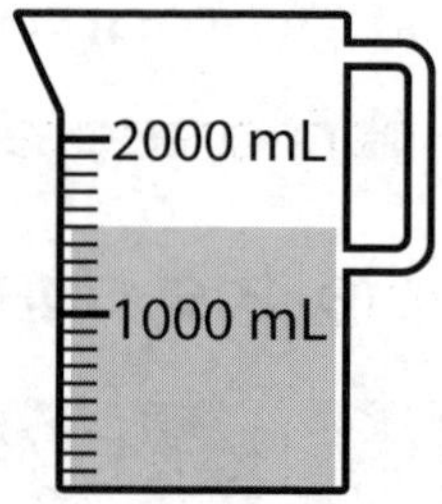

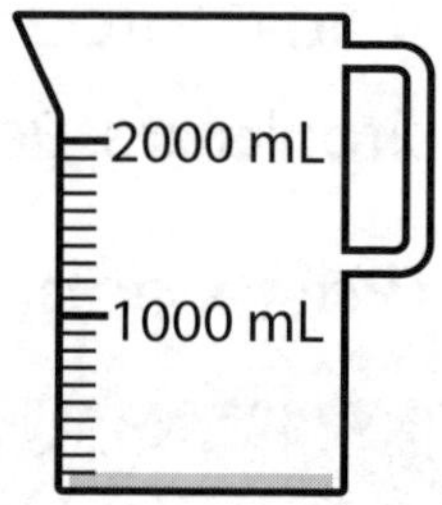

__________ __________ __________ __________

2 Look at the amount of water in this jug.

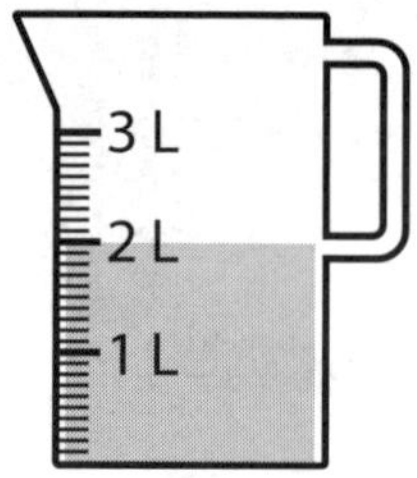

Use a measuring jug to find a container in the room that holds:

a **more** than this.

container: ________________ amount: ________________

b **less** than this.

container: ________________ amount: ________________

c about the **same as** this.

container: ________________ amount: ________________

3 Find and draw 2 containers in the room that are different shapes but have about the same capacity.

Capacity (TRB pp. 138–141)
Using units of measurement Measure, order and compare objects using familiar metric units of length, area, mass and capacity
(VCMMG140)

Unit 30

DATE:

STUDENT ASSESSMENT

1 Name 3 items that are measured in litres (L).

____

2 Name 3 items that are measured in millilitres (mL).

____

3 Number the following items from **smallest** capacity (1) to **largest** capacity (5).

______ ______ ______ ______ ______

4 How many of this container 

would be needed to fill the following containers?

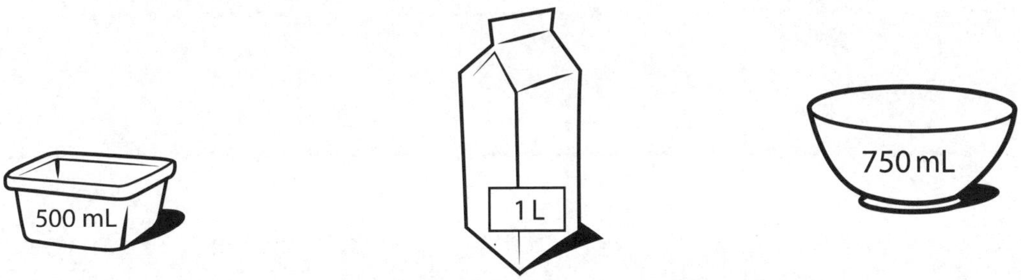

______ ______ ______

5 How much water is in each of these measuring jugs?

______ ______ ______

3D Shapes

DATE:

You will need: a set of 3D shapes

1 Use the 3D shapes to help you complete the table.

	Name	Number of corners	Number of edges	Number of faces	Shape of faces
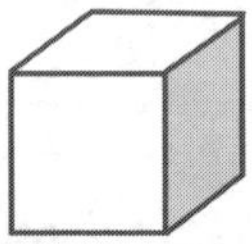					
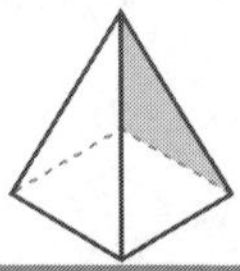					
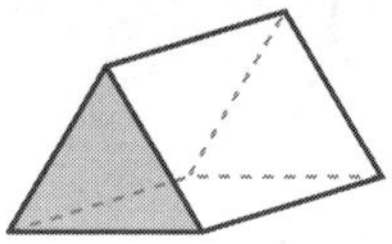					
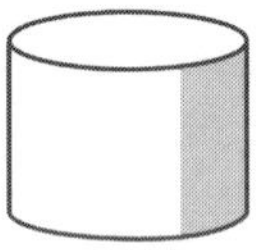					
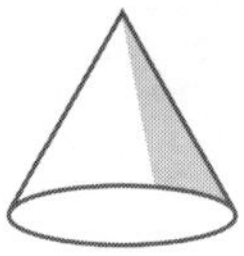					
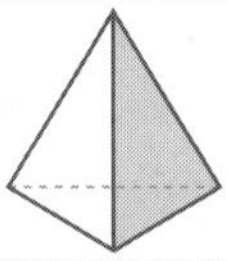					
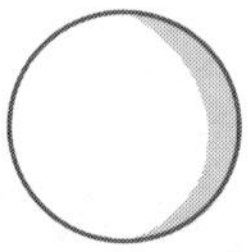					

2 Where have you seen these 3D shapes in everyday places? Draw examples.

cube	rectangular prism
cone	sphere

Pyramids

DATE:

You will need: playdough, straws

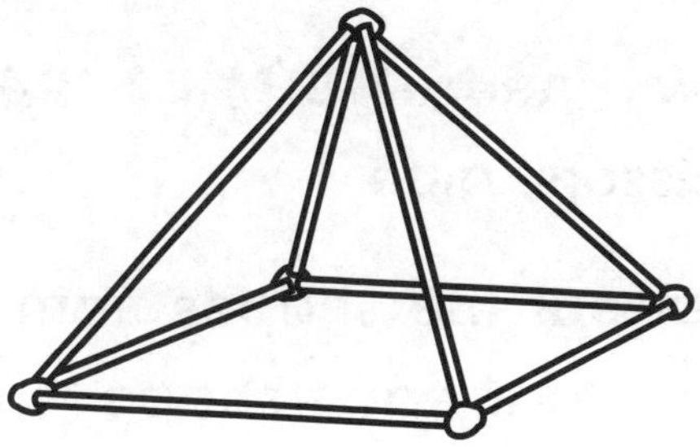

1 Make a model of each type of pyramid using playdough and straws.

2 Have a go at drawing a picture of your model!

3 Write a description of each pyramid. Make sure you describe the faces, edges and corners.

square pyramid	triangular pyramid
hexagonal pyramid	rectangular pyramid

4 What do the pyramids have in common? ______________________

____

5 How are the pyramids different? ______________________

____

6 How are the different kinds of pyramid given their names? __________

____

Prisms and Pyramids

DATE:

You will need: **BLM 86 'Prisms and Pyramids'** (reduced copy), scissors, glue

1 Cut out the shapes from **BLM 86 'Prisms and Pyramids'**.
Decide if each shape is a prism or pyramid.

Paste the shapes on the chart.

Prisms	Pyramids

2 On the chart, draw 2 everyday objects that are prisms. Repeat for pyramids.

3 Look at the shapes in the "Prisms" box. How did you know they were prisms? ______________________________

4 Look at the shapes in the "Pyramids" box. How did you know they were pyramids? ______________________________

DATE:

STUDENT ASSESSMENT

1 Draw a line matching each 3D shape with its description.

I have 6 square faces,
8 corners and 12 edges.

I have 5 corners, 8 edges and
one of my faces is a square.

I have 9 edges and 6 corners.
I have 5 faces.

I have 12 edges and 8 corners.
All of my 6 faces are rectangular.

I have no edges and no corners.
I look like a ball.

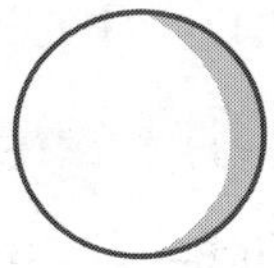
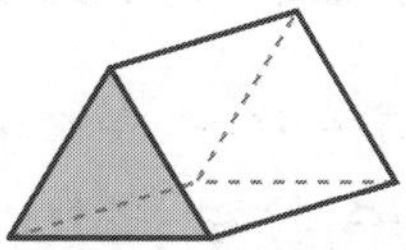
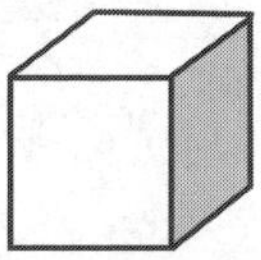
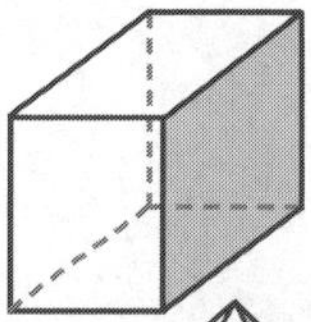
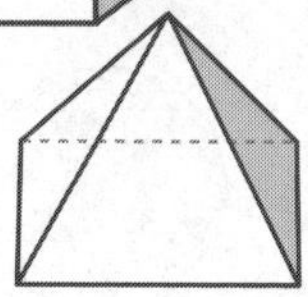

2 Draw the shape of the faces for these 3D shapes.

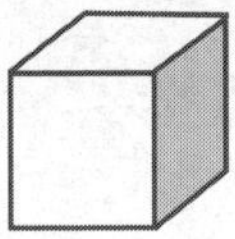	
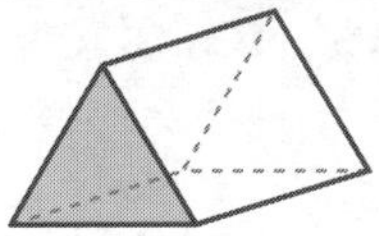	
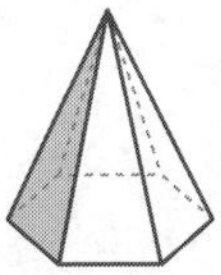	

3 **a** Label the shapes as a **prism** or **pyramid**.

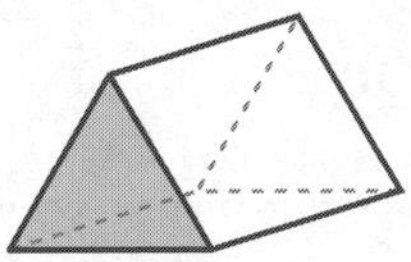
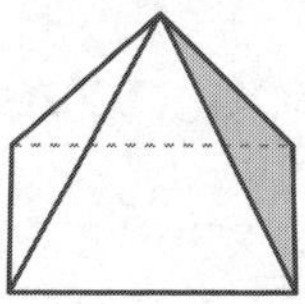
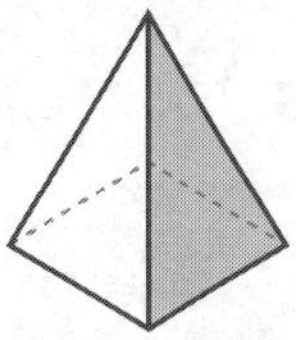
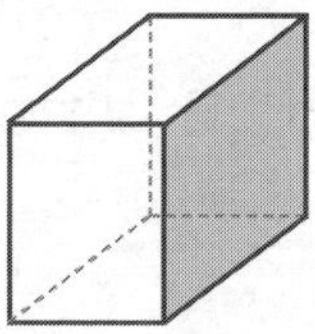

________ ________ ________ ________

b What is the difference between a prism and a pyramid?

____

Coins, Coins, Coins

DATE:

You will need: **BLM 87 'Coins'**, scissors, glue

1 Colour the Australian coins.

2 Write the total value of the following coins.

= ____________

= ____________

= ____________

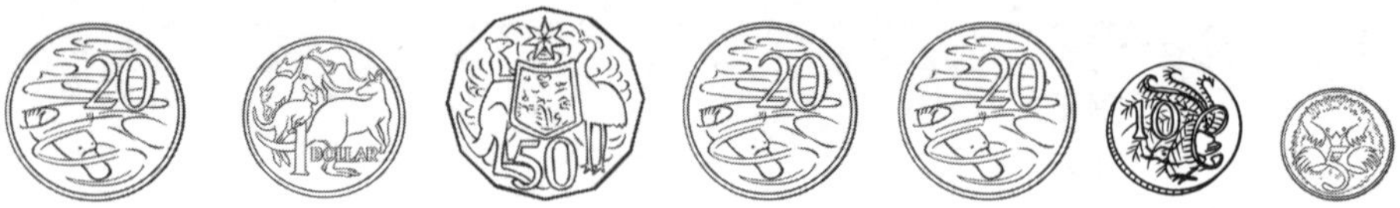

= ____________

3 Cut out coins from **BLM 87 'Coins'** to show:

a 2 different ways to show 40c.

b 2 different ways to make $2 if you have no $2 coin.

Money (TRB pp. 146–149)
Money and financial mathematics Represent money values in multiple ways and count the change required for simple transactions to the nearest five cents
(VCMNA137)

Supermarket Shopping

DATE:

You will need: **BLM 87 'Coins'** and **BLM 89 'Supermarket Catalogue'**, scissors, glue

1 **a** You have $12 worth of coins in your pocket to buy some groceries. Select items from **BLM 89 'Supermarket Catalogue'** to get a total as close to $12 as you can. Paste each item below. Make sure you don't go over $12!

b What was your total amount spent? ________

2 What coins could you use to pay for the items? Present 2 different options by pasting the coins from **BLM 87 'Coins'** in the boxes below.

Calculating Change

DATE:

You will need: **BLM 90 'Canteen List'**

1 Use **BLM 90 'Canteen List'** to work out the change for each lunch order.

a

salad roll

piece of fruit

flavoured milk

$5 enclosed

Change: ______________

b

2 sushi rolls

cheese and biscuits

orange juice

$7 enclosed

Change: ______________

c

vegetable lasagne

plain yoghurt

2 rice crackers

mineral water

$10 enclosed

Change: ______________

d

tuna and salad sandwich

sultanas

fruit jelly

apple juice

$8 enclosed

Change: ______________

2 That day at the canteen, they only have 5c, 10c and $1 coins to give as change. List the change that may have been given for each bag.

Bag **a** ______________________________

Bag **b** ______________________________

Bag **c** ______________________________

Bag **d** ______________________________

Money (TRB pp. 146–149)
Money and financial mathematics Represent money values in multiple ways and count the change required for simple transactions to the nearest five cents
(VCMNA137)

Unit 32

STUDENT ASSESSMENT

DATE:

1 Write the total value of the following coins.

= ________

= ________

= ________

= ________

2 Show 2 different ways to make $1.20 using coins.

3 Show 2 different ways to make $20 using notes and/or coins.

4 Work out the change required for the following purchases from a newsagency.

a Greeting card for $4.95, paid for with a $10 note.
Change: ________

b Wrapping paper for $3.30, paid for with a $5 note.
Change: ________

c Exercise book for $2.40 and a pencil for $2.00, paid for with a $10 note. Change: ________

Find My Rule

DATE:

Find the rule for each function machine and fill in the missing numbers.

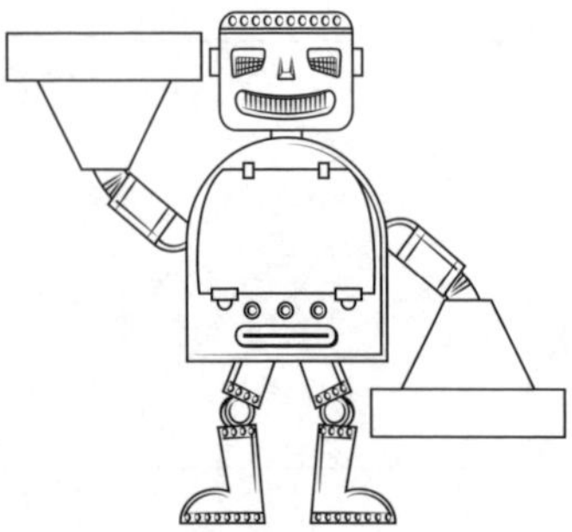

in	out
4	5
10	11
	7
13	

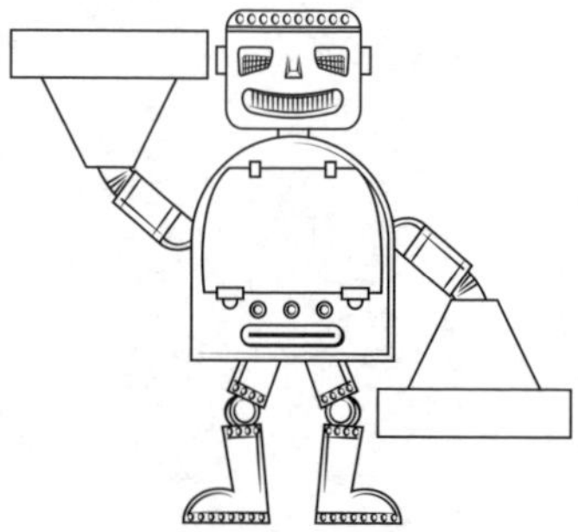

in	out
3	6
2	
7	14
	8

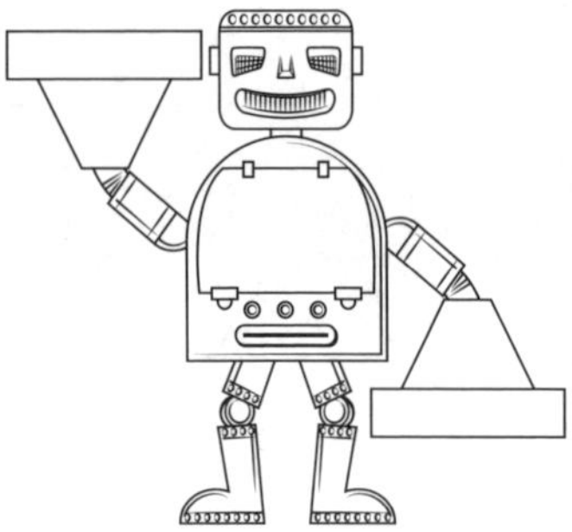

in	out
9	6
	–3
2	
4	1

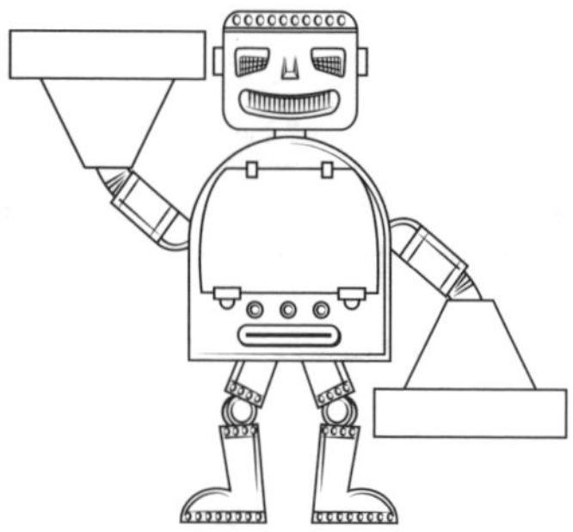

in	out
13	24
7	18
100	
	91

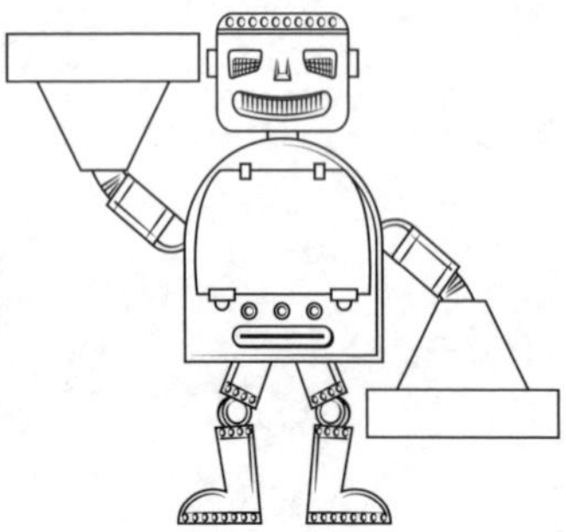

in	out
3	
7	21
	99
8	

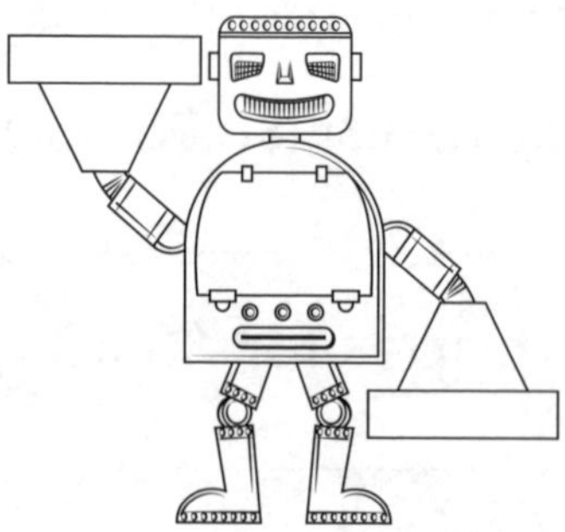

in	out
16	4
	3
80	20
48	

Using Rules – Function and Inverse (TRB pp. 150–153)
Patterns and algebra Use a function machine and the inverse machine as a model to apply mathematical rules to numbers or shape
(VCMNA139)

Reverse!

DATE:

Help Stella find her way home by travelling backwards along the path. Find the inverse to each equation and fill in the empty boxes.

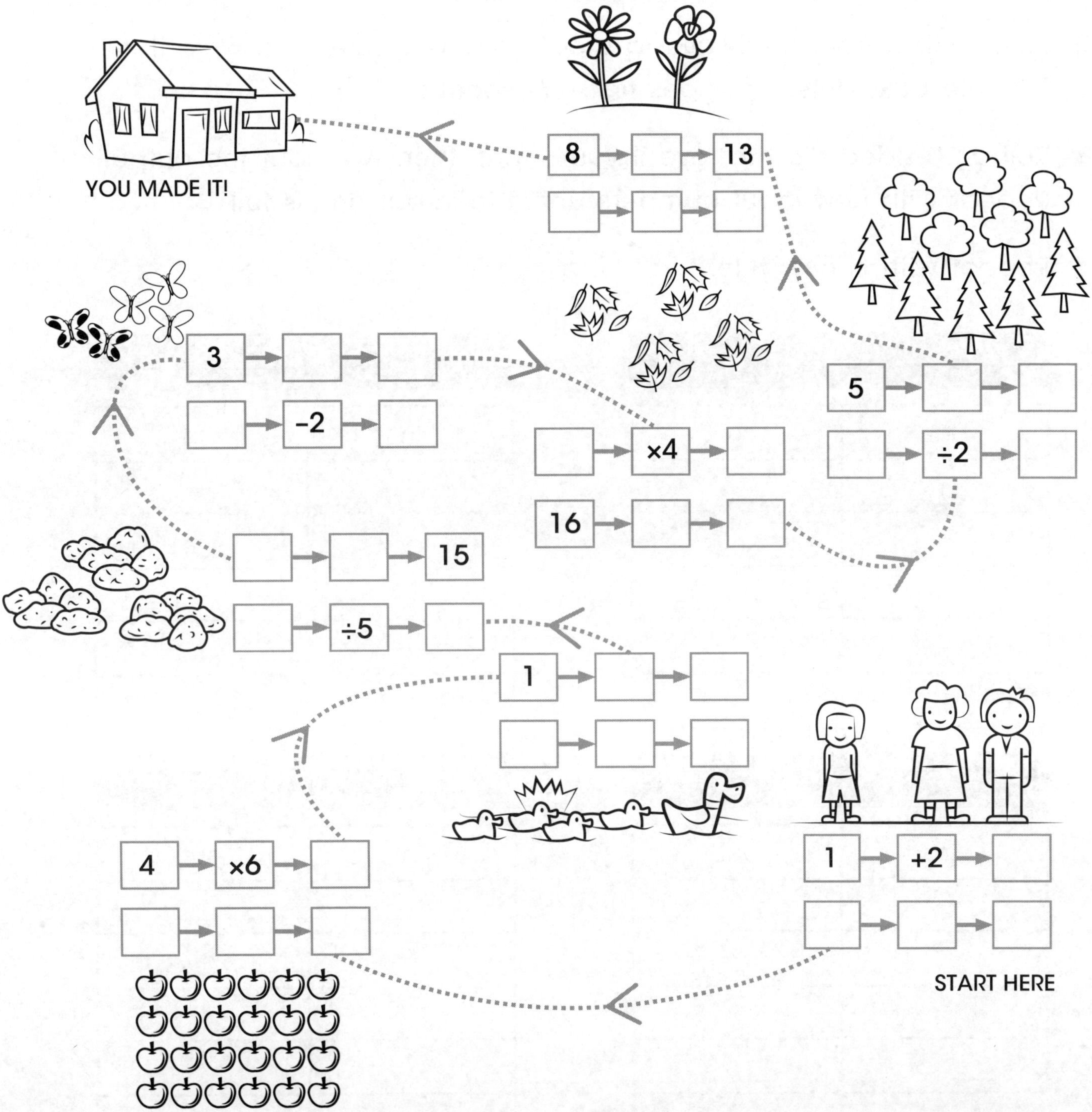

2-Step Rules

DATE:

You will need: **BLM 93 'Function Rule Cards'**, a 10-sided dice

1 Separate the rules cards into two piles: addition/subtraction and multiplication/division.

2 Pick a card from each pile to create a 2-step rule, and write it in the rule box. Note: ×/÷ goes first, +/– second.

3 Roll a 10-sided dice to reveal your input, then work out the output. Repeat with new input numbers until the in-out box is full.

4 Repeat with a new rule.

Rule:

in	out

Rule:

in	out

Rule:

in	out

Rule:

in	out

Unit
33 STUDENT ASSESSMENT

DATE:

1 Find the rule for the following functions:

Rule: ____________________

in	out
3	5
6	8
9	11

Rule: ____________________

in	out
4	–1
7	2
9	4

2 Use the rules to complete these in-out tables:

Rule: × 3

in	out
2	
4	
	33

Rule: + 7

in	out
4	
	15
12	

3 Find the inverse to complete the following:

8	→	+7	→	15
	→		→	

9	→	×3	→	
	→		→	9

4 Use the 2-step rules to complete these in-out tables:

Rule: × 2 – 3

in	out
2	
4	
7	

Rule: ÷ 2 + 6

in	out
8	
12	
22	

Shape Slides and Turns

DATE:

You will need: pattern blocks (or **BLM 98 'Pattern Blocks'**)

1 Use your pattern blocks to complete the table. Trace your responses.

Shape	Slide	Quarter turn	Half turn

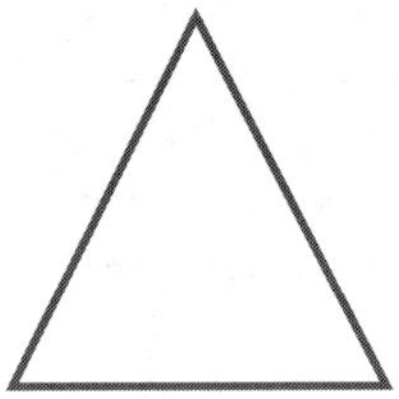			
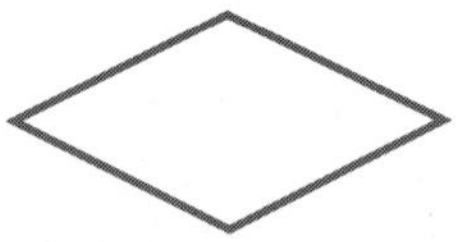			

2 Find a way to arrange your pattern blocks to fill this triangle. Trace your response into the triangle.

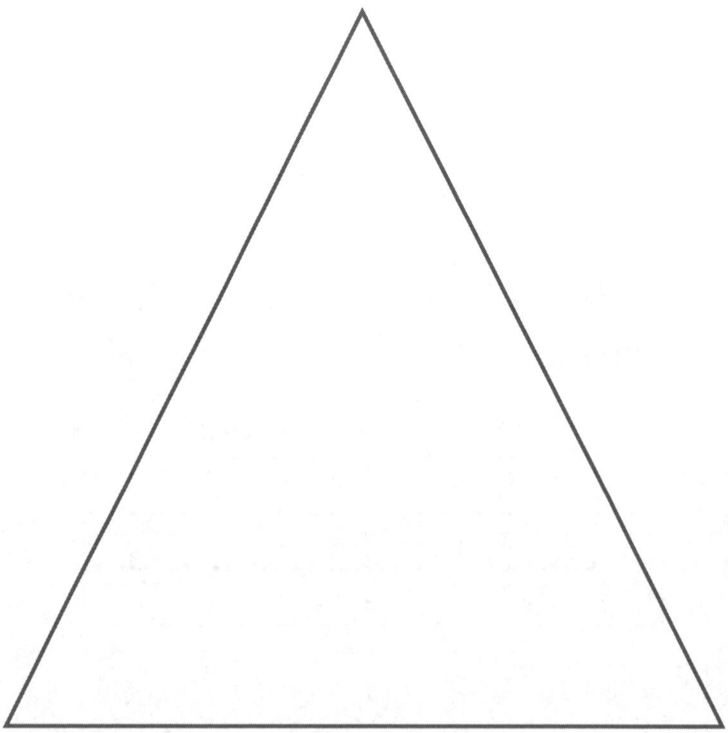

3 How did you use slide and turn to create the triangle?

4 Can you find another way to create the same triangle with the pattern blocks?

Search the House!

DATE:

Colour all of the slides and turns you can find on this house and list them on the Venn diagram.

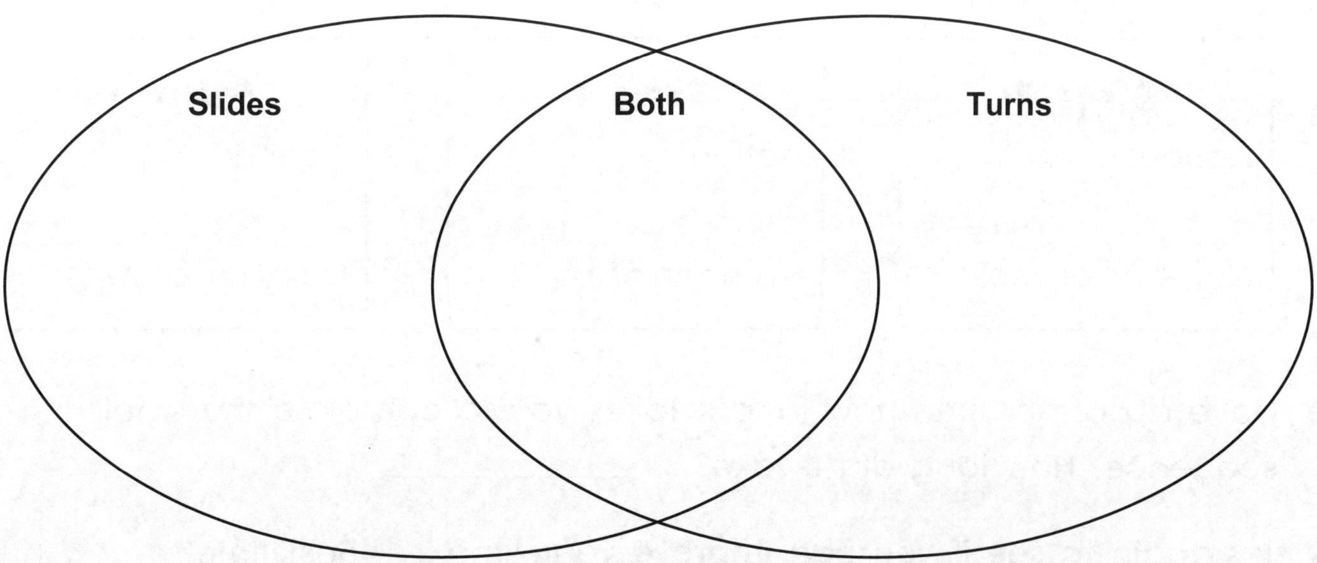

Get Moving!

DATE:

You will need: two different-coloured dice, a stopwatch, a partner

DICE 1 – ACTIONS

1 Slide to the right
2 Slide to the left
3 Quarter turn jump
4 Half turn jump
5 Quarter turn pivot
6 Half turn pivot

DICE 2 – NUMBER OF REPETITIONS

1 Roll the dice to work out your sequence of moves.

Step 1 Action: Number of reps:	**Step 2** Action: Number of reps:	**Step 3** Action: Number of reps:
Step 6 Action: Number of reps:	**Step 5** Action: Number of reps:	**Step 4** Action: Number of reps:
Step 7 Action: Number of reps:	**Step 8** Action: Number of reps:	**Step 9** Action: Number of reps:

2 Have a partner time how long it takes you to complete the whole sequence. How long did it take? ________________

3 Try again to see if you can improve your time. Best time: __________

Unit 34 DATE:

STUDENT ASSESSMENT

1 Complete the table.

Shape	Slide	Quarter turn	Half turn

2 Describe the following in your own words:

Slide: ______________________________

Turn: ______________________________

3 Draw and label two examples of where you may see a **slide** in the real world.

4 Provide an example of where you may see **turns** within sporting activities. Explain how a turn is used.

____

____

____

____

Challenges

Worded Problems

Answer these problems in your maths book or on a sheet of paper.

1 Sarah wrote a 4-digit number. The digit in the thousands place is 4 more than the digit in the hundreds place. The digit in the hundreds place is 6 less than the digit in the tens place. The digit in the ones place is 5 less than the digit in the tens place. The sum of all four digits is 19. What is the 4-digit number?

2 The Year 3 children are going on an excursion. There are 125 students, 6 teachers and 14 parents riding in the bus to the aquarium. If a bus holds 40 people:

a How many buses will they need?

b How many spare seats are there?

3 I went to the park riding my new bicycle. At the park there were 15 other bicycles and tricycles. There were a total of 38 wheels. How many tricycles were there?

4 There are 24 Boy Scouts marching in the parade. They have to march in equal rows and each row must have more than one person. Show the different ways they could arrange themselves to march in the parade.

5 Jack had a list of digits: 9 8 7 6 5 4 3 2 1. He put addition signs between some of the digits and then added the numbers together. How did he make a total of 99 if:

a He used 6 addition signs?

b He used 7 addition signs?

6 Luke's café has 3 tables. The smallest table has 4 less seats than the medium table, and the largest table seats twice as many as the medium table. One busy afternoon, three-quarters of the seats in the café were taken. When 8 more people arrived, Luke was only able to seat 5 of them.

a How many seats are there in the café?

b How many seats are there at each table?

7 Lucy, Maya and Freddie are all putting in money to buy a gift for Su-Ling's birthday. They buy her a new diary for $12.95. Maya contributes $4.00, and Lucy puts in 55c more than Freddie. How much money do Lucy and Freddie each contribute?

Decimal Puzzles

Work out the following decimal equations and place the letters above the answers in the puzzle to find out:

What happened when the teacher told Max a joke about decimals?

D 0.5 + 3.5 =

N half of 4.6 =

E 9.6 − 4.3 =

O half of 12.4 =

G 3.2 + 4.4 − 0.2 =

P 4.4 × 2 =

H 4.4 − 4.3 + 1.1 =

T 2.1 × 3 =

I 5.5 + 3.1 − 4 =

1.2	5.3		4	4.6	4	2.3		6.3
7.4	5.3	6.3		6.3	1.2	5.3		
					!			
8.8	6.2	4.6	2.3	6.3				

Magic Squares

Using the numbers 1–9, make each row and column add up to **15**.

Using the numbers 0–15, make each row and column add up to **30**.

Answer these problems in your maths book or on a sheet of paper.

What Could I Wear?

Peter has a yellow shirt, a blue shirt, brown shorts, green shorts, black shoes and red shoes. How many different outfits could he wear?

Symmetrical Flowers

Lily is busy in the garden. She has 4 plants with pink flowers and 4 plants with white flowers. She is planting them in a single row in a planter box. She wants to plant them so the flowers are symmetrical in the box. Show the different ways she could plant the flowers.

Shape Hunt

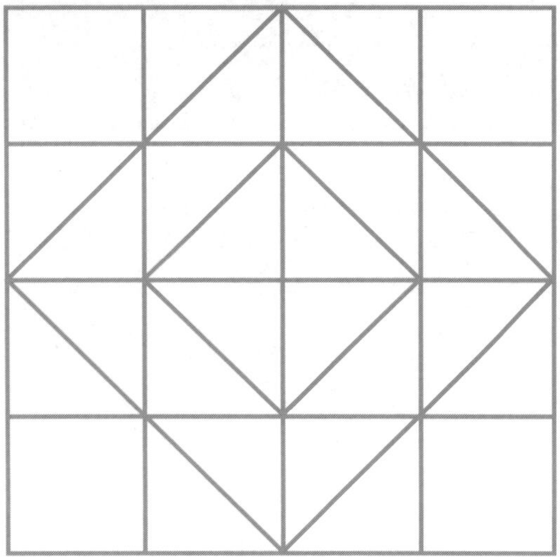

Look carefully at this design and count the shapes.

How many squares?

How many triangles?

What other 2D shapes can you see?

How many of these shapes can you find?

Matchstick Problem

12 matchsticks have been arranged as shown.

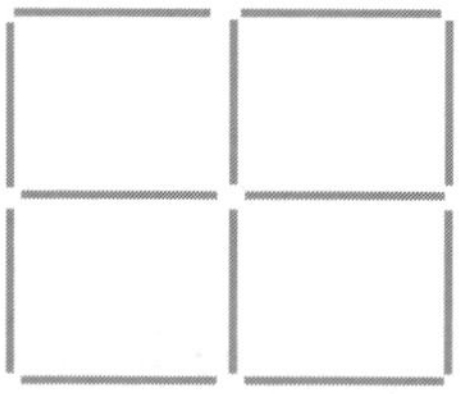

Can you move 2 matchsticks to make 7 squares?

Can you move 3 matchsticks to make 3 squares?

Can you move 4 matchsticks to make 3 squares?

Maths Glossary

4 Operations

terms and signs

- **+** addition
- **-** subtraction
- **×** groups of (multiplication)
- **÷** shared between (division)
- **=** (equals)

array
an arrangement of objects in columns and rows

difference
the answer to a subtraction equation

$10 - 6 = 4$ ← difference

divisor
a number divided into another number

$18 \div 6 = 3$ ← divisor (6)

product
the answer to a multiplication equation

$2 \times 3 = 6$ ← product

sum
the answer or total to an addition equation

$4 + 5 = 9$ ← sum

total
the sum or answer to an addition equation

$4 + 5 = 9$ ← total

quotient
the answer to a division problem

$8 \div 2 = 4$ ← quotient

remainder
the amount left over when solving a division problem

even
a whole number that is divisible by 2

odd
a whole number that is not divisible by 2

strategy
the plan or steps used to solve a problem

calculate
to work out the answer

Rounding

rounding
to change the value of a number to make it easier to estimate. Numbers can be rounded to the nearest 5, 10, 100, 1000, etc.

rounding up
Numbers 5-9 are rounded up.

rounding down
Numbers 1-4 are rounded down.

Money

5c

10c

20c

50c

$1

$2

Fractions and Decimals

$\frac{1}{2}$	$\frac{1}{4}$	$\frac{1}{5}$	$\frac{3}{4}$	$\frac{1}{3}$	$\frac{2}{3}$
half	quarter	one fifth	three quarters	one third	two thirds
0.5	0.25	0.2	0.75		

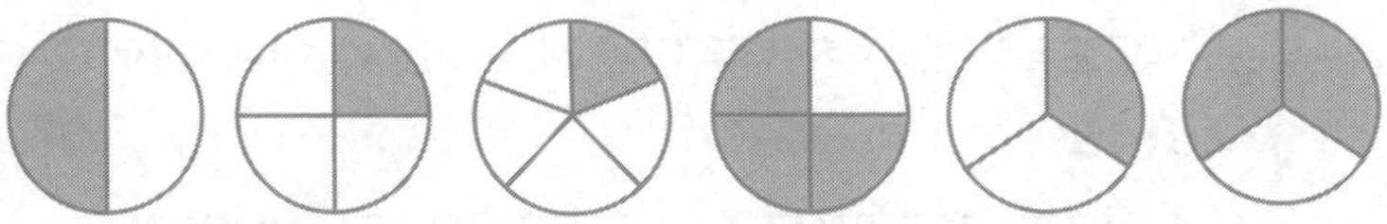

denominator
the bottom number in a fraction. It shows how many parts there are in a whole

$\frac{4}{5}$ ← denominator (5)

numerator
the top number in a fraction. It tells how many parts of a whole there are

$\frac{4}{5}$ ← numerator (4)

decimal point
the point that separates a whole number from a decimal fraction

0.5 ← decimal point

Maths Glossary

Units of Measurement

analogue clock

a clock with two hands and 12 numerals on its face. It is used for showing the time

digital clock

a clock that uses only numbers to show the time

mm	millimetre	**g**	gram
cm	centimetre	**kg**	kilogram
m	metre	**mL**	millilitre
m^2	square metre	**L**	litre

Angles and Lines

curved line

a line that is not straight

diagonal line

a straight line that is not a horizontal or a vertical line

horizontal line

a straight line that is parallel to the horizon

vertical line

a straight line that is perpendicular (a quarter turn) to the horizon

angle

a measure of turn

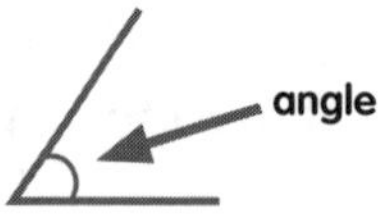

Location

directions

instructions to get from one place to another

map

a simplified bird's-eye-view drawing of a particular location

coordinates

a pair of numbers or letters that show a particular point on a map or grid

Shape

two-dimensional shape (2D)

a shape that can be cut out of paper. 2D shapes have only 2 dimensions: length and width

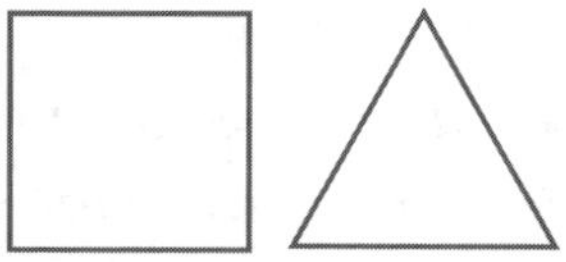

three-dimensional shape (3D)

a solid shape with 3 dimensions: length, width and height

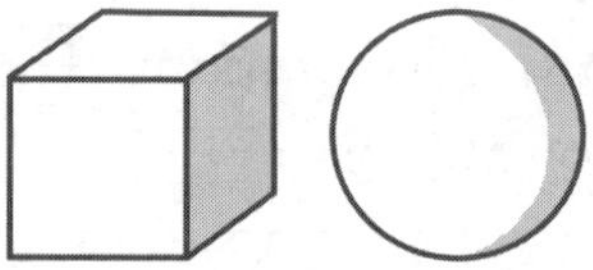

pyramid

a solid shape with a polygon base (e.g. a square) and triangular faces

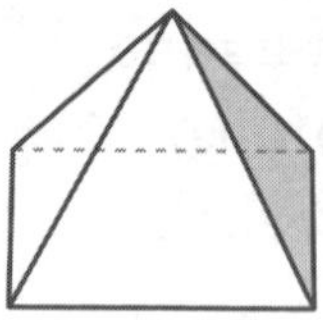

prism

a solid shape with two identical polygon faces that are parallel, and parallelograms (including rectangles) for all other faces

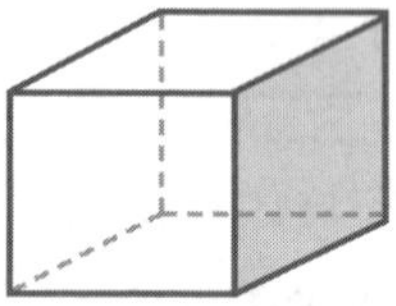

symmetry

a shape has symmetry if both its parts match when folded along a line

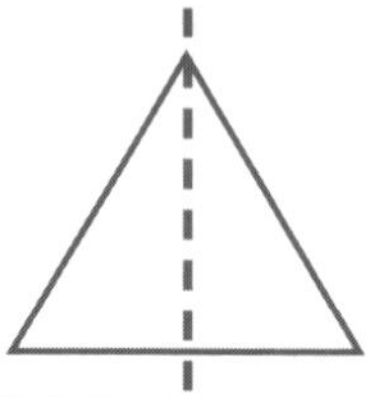

Statistics and Probability

prediction

a guess about the likelihood of an event happening, based on existing information

data

factual information gathered for further discussion

survey

a question or questions answered by a group of people

graph

a visual representation of data